PRAISE FOR *FACEHOOKED*

"The biggest change in modern romance is Facebook; it not only allows old flames to reignite, but it is also increasingly cited as a top reason for divorce. Used by an out-of-control government to evaluate the mood of the nation, Facebook is more than just what your kids had for dinner. Like Oz unveiled, *Facehooked* reveals the truth behind the screen."

—Mancow Muller, syndicated radio and Fox News Channel TV host and best-selling author of *Dad, Dames, Demons, and a Dwarf: My Trip Down Freedom Road*

"*Facehooked* talks about how teens connect through social-media networks like Facebook, Twitter, Snapchat, and YouTube. Now more than ever, teens have a voice in the world and can even make a serious social impact! The only problem is—some people get into trouble when they share too much. Kids seem to feel obligated and addicted to responding and uploading to Facebook when it's not really necessary; doing so just adds more pressure and raises the potential of posting things they normally wouldn't or shouldn't. Whether online or logged off, I think people can express themselves any way they want as long as they treat each other with respect and stay true to themselves."

—Sean Giambrone, Actor, *The Goldbergs*

T0294293

"When it comes to understanding the world of teen online interaction, Dr. Flores gets it. *Facehooked* sums up what I've suspected is happening with today's teens. Dr. Flores provides a valuable wealth of information regarding smartphone and social-media addiction, digital expression, and the negative (and positive) effects of sharing personal information through various social networks. She clearly explains the influence social media can have on teen expression and self-perception, and provides clear guidelines for parents on how they can protect their teens' privacy. *Facehooked* is an intelligent, comfortable, and important read for any student, parent, school administrator and policy-maker who wishes to understand how social media is shaping today's 'Digital Natives.'"

—Dr. Matthew Clark, Child and Adolescent
Psychologist, Director of The Clark Institute

"Dr. Flores offers a compelling and genius look at what is really going on when seeking that Facebook fix. Is your hidden agenda personal validation? Creating a fantasy life? Or are you actually addicted? Her explanations are eye-opening and will make you rethink how and why you interact on Facebook every time you log on."

—Shawne Duperon, Six Time EMMY®
winner and Founder of Project: Forgive

"A fresh voice for psychology, Dr. Suzana Flores is in touch with the newest trends in communication and interaction. *Facehooked* is a must-read for anyone who's on Facebook and other social networks."

—Jose Jara, Founder, LatinoScoop.com

"Dr. Suzana Flores's book is a must-read. It shows us how we are getting away from self-love and true soul connections with others and veering into a digital and more superficial sense of love, self and connection. Our goal in life should be to love in person, to love ourselves and to return to soul-connected relationships."

—Sherrie Campbell, PhD, author of
Loving Yourself: The Mastery of Being Your Own Person

"Facebook connects more than one-seventh of the planet. It's an honorable feat. But it's increasingly uniting our culture by way of unsteady anxiety and greed. Dr. Flores examines the roots of our social-media addiction with nightmarishly relatable honesty. Readers will leave with practical tools for change, and the ironic knowledge that we're all experiencing Facebook's emotional cataclysm together."

—Stephanie Buck, Features Editor, *Mashable*

"*Facehooked* is a provocative analysis of how social media has impacted the social structure of human relationships, self-awareness and behavioral addictions. Author Dr. Suzana Flores has been able to humanize the issue for the reader through real-life scenarios that highlight the depth of impact Facebook can have on a personal and social level. This read is one that every smartphone and social-media user can identify with, and should heed as a cautionary reminder of our own behaviors."

—Christina Montes Scott, Director, Chicago Latino
Book & Family Festival & Publisher for *TeleGuia Magazine*.

"Facebook is a wonderful and terrible imaginary yet real social world that powerfully affects its users. In this book, Dr. Suzana Flores offers profound insights about Facebook and regales readers with real-life Facebook stories . . . stories that are simultaneously shocking and routine. If you've ever felt—and then given in to—the impulse to check your Facebook page, this book is for you. In fact, you should try this experiment: Don't check your Facebook page again until you start reading this book. I wonder if you can?"

—John Sommers-Flanagan, PhD, Professor of Counselor Education, University of Montana, Author of *Tough Kids, Cool Counseling* and *Clinical Interviewing*

"We live in an era where disruptive technology waves are constantly crashing on our shoes. There are those among us who will ride these waves to enhance their lives, and the lives of those around them. And there are those who will allow these waves to crush them, and disrupt their lives. As a parent and a professor, I strive to teach my kids and students to avoid being the latter by understanding what type of problems each technology brings with it. *Facehooked* is a great introduction into the world of technology and a fascinating read for those who wish to understand some of the problems that may come with technology and social media."

—Faisal Akkawi, PhD, Executive Director, Information Systems Programs, Northwestern University

"Replete with interesting examples and real-life stories, *Facehooked* explores the good, the bad, and the ugly aspects of our fascination with social media and provides practical and helpful advice for healthier connections in our increasingly electronic social world."

—Dr. Sherrie Bourg Carter, Author, *High Octane Women: How Super Achievers Can Avoid Burnout*

"When I began reading *Facehooked*, I couldn't put it down! The timing of this publication and the contribution of her social research could not be more relevant on how much of an impact social media has on our everyday lives. Whether one sees it as a blessing in disguise, a therapeutic fix, as something intrusive or addictive, Facebook is here to stay. We are all, whether we like it or not, *Facehooked*!"

—Joseph A. Davis, PhD, author of *Stalking Crimes and Victim Protection: Prevention, Intervention, Threat Assessment, and Case Management*, and Contributing Author at *Psychology Today* magazine

"Dr. Flores brings to light how digital communication is impacting us in ways we never predicted. *Facehooked* is a social analysis of our personal and social disconnection in a globally connected world, and a call for us to return to reality...and each other."

—Bryant McGill, Author and Entrepreneur, McGill International

"If you've ever checked a 'Relationship' status, argued over a post, felt jealous, compulsively checked Facebook, or wanted a life that matched your profile, then *Facehooked* is a must-read. Dr. Flores provides many engaging vignettes from her years' long study of America's favorite time sink that explain why Facebook is so compelling, yet also distracting, anxiety provoking, and sometimes emotionally risky for individuals and relationships alike. Finally, you've got a real friend in Facebook—it's *Facehooked*!"

—Dr. B. Janet Hibbs, Psychologist and Author of *Try to See It My Way: Being Fair in Love and Marriage*

"*Facehooked* expertly exposes the addiction that social media can become. A must-read for anyone who wonders why they spend so much time on Facebook and what they're getting out of it."

—Ron Kelly, Talk Show Host, *The Ron Kelly Show*

"*Facehooked* is an incisive analysis detailing the current impact of social media on our lives. Dr. Flores's insightful and compelling work offers a fresh perspective to a phenomenon that continues to evolve. Flores moves beyond the 'culture' of social media to explore the insidious psychological ramifications facing our society. Her message is ultimately a hopeful one—that strategies to increase mindfulness will free us from Facebook addiction. This book will change the way we view social media, as well as ourselves."

—Sarah Suzuki, LCSW, CADC,
Author of *Imperfect Circles*

"In *Facehooked*, Dr. Flores has drawn from her many years of clinical work to tell the story of how social-networking outlets can transform one's genuine self-concept into that of a virtual identity. Through interesting, compelling, and at times tragic patient vignettes, Dr. Flores brings to life a type of addictive behavior that can have the same destructive impact as alcohol, gambling, or serious obsessive-compulsive behaviors. This is an important resource for those providing counseling to individuals whose lives have been transformed by a technology that allows a computerized self to feel more genuine than one's actual self."

—Thomas Gettelman, PhD, Vice-
President, Carolinas Healthcare System

"Dr. Flores's first book, *Facehooked*, is one of the most detailed and significant works on this very timely subject. Her professional evaluation of the impact of Facebook on our society resonates particularly with my experience working with traditional and non-traditional students in higher education over the last 20 years. Since the issue of finding significance in the reflection of others' opinions can be so compelling for some, the in-depth investigation into one of our society's most pervasive venues is not only appropriate but absolutely needed!"

—Valarie Rand, MEd, Associate Dean of Student Affairs, The Illinois Institute of Art-Chicago

"No longer will you view Facebook as a benign pastime. The real-life examples offered by Dr. Flores provide some eye-opening consequences of Facebook use. It is a cautionary tale to all of us that work with users of this larger-than-life social medium. The parallels to drug and alcohol use are spot-on and thought provoking. Dr. Flores shows us how we can use the technology in a responsible way, which is much-needed advice."

—Caryn Feldman, PhD, Clinical Psychologist and Assistant Professor of Physical Medicine and Rehabilitation

"An interesting answer to an interesting question: Facebook can have a real effect, sometimes quite detrimental, on how people view themselves—and not just virtually. How our virtual identities affect our lives in the world is a question everyone on Facebook needs to ask themselves."

—Eric Brinkman, Author of *Easy Tibetan Logic*

"*Facehooked: How Facebook Affects Our Emotions, Relationships, and Lives,* could not be more timely. As more of us connect up online through social networks and share our lives with varying degrees of openness, it's good to have our eyes wide open about the impact and influence on ourselves, our connections, and our world. Facebook has real potential for community building and working together for positive change, but it also has traps, pitfalls and stumbling blocks. And online, bad behavior takes many frustrating forms. Understanding what we're getting into with Facebook is an important and much-needed conversation, and Dr. Suzana Flores has given us more than enough examples, stories and ideas to get it started. I was fascinated with her insights from page one and simply couldn't put it down."

—Dr. Rick Kirschner, Author, *How To Click With People* Coauthor, *Dealing With People You Can't Stand*

"*Facehooked...* Dr. Flores explores the impacts of social-media addictions for those who view their life's value through it. She describes the phenomena of social-media compulsion and how this addiction has shaped our relationships. This is a must-read for school administrators, law enforcement and parents trying to understand the implications of these compulsions. As social media reaches more deeply into all of our lives, it will shape and drive many of our actions and thoughts in the future; this work reviews the pitfalls of this dependence."

—Chief Patrick J. O'Connor, CPC, President: Illinois Campus Law Enforcement Administrators Association

"Dr. Suzana Flores has written an impactful book that is not only necessary, but also critical to understanding the impact of Facebook on our sense of self, our changing culture and most importantly on the fate of our young people. Through poignant stories of real people she has revealed the depth of this international phenomenon in a manner that allows us to see below the surface and into the core of the effects of this social-media tool. We cannot turn back the clock. Facebook is with us. But through Dr. Flores's insights we can understand the consequences of being addicted and consumed by an Internet tool that has hurt many in ways few have understood prior to the wisdom contained in this excellent book. Highly recommended."

—Arthur P. Ciaramicoli, EdD, PhD, Chief Medical Officer, Soundmindz.org, Author of *The Power of Empathy, Performance Addiction,* and *The Curse of the Capable*

With her trademark humor, Dr. Flores is able to speak intelligently about a topic in which we have become consumed and overly dependent upon, yet she avoids the overuse of 'jargonistic' hyperbole often associated with academics. Use of social media has very much become a part of our daily lives. Her witty, smart and practical approach allows a wider range of readers to understand why we are so 'hooked' on social media and other virtual methods of socializing."

—D. Dontae Latson, MSSA, President & CEO, YWCA McLean County

"*Facehooked* brings to light the psychological stressors surrounding Facebook/social media and a person's ability to change themselves—within their own mind. As a clinician who treats individuals who struggle with recognizing their problems and deficits, *Facehooked* brings a new awareness about how people are hiding, changing, and denying their authentic selves, while creating a false persona in order to hide their insecurities. I particularly like how, at the end of each chapter, Dr. Flores offers insight and guidance on the subtle changes individuals can make that can lead them to feel more secure with who they are and how they choose to express themselves on this very public platform."

—Dr. James W. Atchison, Board Certified Specialist
Physical Medicine and Rehabilitation and Pain Medicine

"As a media observer and educator, I find this discussion of emerging problems in our social-media-saturated reality insightful and necessary if we are to understand how pervasive and possibly negative the impact of digital communication can be on our lives. Of particular interest to me is how instant access to so much information about others and the equally unbridled sharing of information from one person to so many is affecting Millennials, many of whom may have grown up without knowing about or experiencing a reality devoid of social media and its influence on their behavior. They may very well ask, "What's the problem?" In her book, *Facehooked*, Dr. Flores sheds critical light on what is happening to us as users and probable abusers of social media."

—Joseph Berlanga, MFA Adjunct Professor,
Culture, Race and Media School of
Media Arts, Columbia College Chicago

facehooked

facehooked

*How Facebook Affects Our
Emotions, Relationships, and Lives*

Dr. Suzana E. Flores

REPUTATION BOOKS

FACEHOOKED

Published by Reputation Books
reputationbooksonline.com

Copyright © 2014 by Dr. Suzana E. Flores
All rights reserved.

No part of this book may be used or reproduced in any manner whatsoever
without written permission from Reputation Books, except in the case of brief
quotations embodied in critical articles and reviews. For information, contact
publisher at reputationbooks@gmail.com.

Cover Image: Richard Baukovic

Book Design: Mary C. Moore and Lisa Abellera

Library of Congress Cataloging-In Publication Data (TK)

ISBN 978-1-944387-33-4 (hardcover)

ISBN 978-1-944387-34-1 (trade paper)

ISBN 978-1-944387-35-8 (e-book)

Second Edition: October 2021

10 9 8 7 6 5 4 3 2 1

Reputation Books

Dedicated to all my Facebook friends who
wonder if I'm writing about them.
I am.

CONTENTS

FOREWORD

WHEN TECHNOLOGICAL INNOVATION OUTPACES SOCIAL innovation (which is nearly always), the results can be both exciting and frightening. Social media has impacted us in ways we could have never imagined. Our relationships, our perceptions of ourselves and others, and what we value and believe we need have all been directly affected by our experiences on Facebook, and Twitter, and we will be affected by every subsequent instance of immediate, disseminated and disruptive social-media technology. The positive impact of this evolution cannot be underestimated. We are now capable of innovation at a staggering pace, and our ability to educate and communicate is nothing short of miraculous. The resulting Internet-driven transformation of society is now so widely understood and accepted that it hardly makes for news anymore. But our personal metamorphoses—the profound alterations of our relationships with others and ourselves—are not nearly as well understood.

Until now.

As an addiction psychiatrist and former medical director of some of the nation's oldest and largest addiction treatment centers, I'm well acquainted with the impact that addictive behaviors can have on our personal and professional lives. I've seen lives destroyed by alcoholism and drug addiction, food and exercise addiction, gambling and gaming addictions and many other self-destructive behaviors. I've seen firsthand how genetics and environment interplay to create and sustain addiction, and destroy the lives of the sufferer and anyone in their path or wake. So many cases have been heartbreaking. Often the most heartbreaking feature is that these conditions are treatable, but stigma and access prevent people from getting the help they so desperately need. The hope that is present in the treatability of these conditions is destroyed by shame, stigma, and inability to access help.

More recently, we've seen a new wave of addictive behaviors—those that relate to the compulsive use of social media. There is no doubt: we're *hooked*. The consequences of our online behaviors can be no less devastating than for other addictions. The accessibility and ease of use of social media (apps on the phone that's in your pocket or integrated with your car) make these behaviors particularly difficult to stop. The emerging frontier of the Internet of Things will only magnify the impact of social media on our perceptions, mental well-being and relationships.

To be sure, the potential favorable effects of technology on our personal growth are extraordinary. But unchecked and unguided, these same innovations can also accelerate loneliness and despair, and destroy intimacy and emotional connectedness. Take, for example, the importance of privacy in a healthy human experience. We can't grow without the safety it affords. Yet the expectation of online privacy is both exceedingly common and astonishingly naïve. Similarly, the atrophy of our social skills that accompanies avoidance of *real* relationships can be difficult to recover from.

In *Facehooked*, psychologist Dr. Suzana Flores masterfully explores the implications of this emerging universal virtual connectedness on all aspects of our self-perception, expectations, needs and relationships. Because clinical research on these effects is in its infancy, Dr. Flores walks us through personal examples and powerful, touching stories, all the while sharing perceptive insights that help the reader grasp the myriad implications of Facebook and social media on our psychological and social well-being. And unlike much that's been written on this topic, Dr. Flores also explores the *positive* effects that healthy social-media use can produce. Noting that, "Facebook is not the problem," she takes readers on a case-driven exploration of what drives the distressful consequences of maladaptive social-media use and how to avoid them.

Her writing is easy and comfortable, and the journey through her experiences with patients is fascinating and compelling. Even as an experienced addiction psychiatrist, I often found myself saying, "Wow, I've never thought of it that way before." Whether it's FOMO (fear of missing out), the rules and implications of "unfriending," the ever-present risk of cyberbullying, or even more subtle phenomena like our need for Facebook-validation, Dr. Flores covers it, and covers it well.

This extraordinary window into the lives of people from all walks of life affected by social media would itself be worth our attention. However, *Facehooked* is not merely an explanation of the impact of social media on our well-being. Dr. Flores also walks us through what to do about it, so we have the best chance of benefitting from the enormous potential of social media without being buried emotionally and socially by it. The suggestions she offers can really help users "check-in without checking out." Future generations will view Dr. Flores's work as truly prescient, as we all continue our struggle to evolve socially as well as technologically.

Facehooked is a must-read for anyone affected by social media, which is everyone.

Omar Manejwala, MD
Author of *Craving: Why We
Can't Seem to Get Enough*
(Hazelden Publishing 2013)
Charlotte, NC

INTRODUCTION

HOW IS IT EVEN POSSIBLE THAT FACEBOOK CAN SEND someone to the Emergency Room? When I admitted a client to the Psychiatric Unit, after he saw something on Facebook that sent him into a tailspin, I was rattled. But I realized the impact of Facebook was something I'd been seeing more frequently with my therapy clients.

One client missed two sessions in a row after her friends posted comments about her weight gain after having seen recent photographs of her. Another was experiencing problems with his girlfriend's suggestive and flirtatious comments to other male "friends" on Facebook. When I asked my colleagues if their clients were experiencing similar issues, many said yes. Clients were revealing their personal and interpersonal problems now had a new dimension: Facebook.

Many people feel that having no social network is the same as having no place to socialize with friends. Socializing with others on Facebook has become much more than a passive form of entertainment—for many, it's a legitimate way to express ourselves and to record almost every moment of our lives. Many of us take our

Facebook profiles pretty seriously. So seriously in fact that inter-acting on Facebook has caused us to experience a new kind of existential crisis: "If I do something and I don't post it on Facebook, did it ever really happen?" Sure, you can connect with friends and family found across the globe, discover career opportunities and keep a digital account of all your achievements and relationships. But what Facebook users often don't take into account are the many ways in which Facebook can harm our relationships instead of improve them.

Little by little, I began noticing a significant shift in the importance we were placing on our social-network interactions. Facebook posts triggered people to react—often stirring up strong emotions—to what they see (or think they see) on their computer screens. I noticed people making assumptions about other people's Facebook posts and projecting their own feelings onto what such posts "really" meant. People were becoming paranoid, "Was that about me? Is she insulting my kid? He just checked into a club—is he trying to make me jealous? Is she trying to tell me something by posting that song?" Sometimes they were right, sometimes they were way off, but either way, they weren't talking to each other anymore. Instead of checking-in and asking what a post meant, people often made their own assumptions based on a small tidbit of information on their News Feed. They often impulsively reacted by posting negative comments, sharing passive-aggressive posts, or distancing themselves from their friends, many times without just cause. It was clear to me then that Facebook had far more power over people's emotional reactions than we'd like to believe—and it was messing many of us up, one Facebook post at a time.

Facebook has also caused our self-worth to become defined by how many "likes" we receive, our "relationship status," our carefully edited photographs—"Look at how happy I look with my new girlfriend, my new car, my new outfit." Most of us post our celebrity moments and filter what we really share—perhaps avoiding sharing our real selves. In many ways, we have started to connect and interact through this alternative and carefully created fantasy world and it's affecting our real-life emotions.

I've often wondered if Facebook is causing some people so much pain, why not simply log off? One reason. People are addicted. Our sense of self is now being shaped through what we share and how we share it. Through our Facebook friends' likes, shares and comments, we are receiving new messages about what is acceptable, what we should do and who we should be. Such public endorsement is intoxicating and it's leading to addictive behaviors such as checking and rechecking our News Feeds and updates several times a day. Facebook is the ultimate time suck, yet many of us don't realize just how hooked we are until we try to spend 48 hours completely unplugged.

Have our online personas started to take precedence over our offline worlds? I wondered if I was alone with my concern over the long-term effects of social-media interaction. I wanted to speak to people about their own experiences on Facebook but didn't want to just meet with people who thought or felt like I did, so I came up with an idea. I made a huge sign that read, "TALK TO ME ABOUT SOCIAL MEDIA" and held it outside my office in downtown Chicago and patiently waited for people to start approaching me.

Standing outside alone, holding up my big sign on a busy sidewalk in the freezing Chicago weather, I certainly received a lot of "you're crazy" stares from strangers who passed me by, but I didn't care—I was on a mission. One by one, people slowly started to approach me with their stories. With their permission, I recorded many of their accounts with my trusty digital recorder. On some occasions I had an entire group of people discussing how Facebook has affected their self-esteem, their work productivity, their relationships and friendships, and the way they perceive themselves and others. I invited people to meet with me individually to share their more personal and intimate accounts of how Facebook interaction has affected them.

I opened up my questions to the Facebook audience, inviting people from across the globe to tell me their stories. I interviewed teenagers, mothers, doctors, teachers, students, other psychologists, child psychologists, Internet and social-network specialists—all enthusiastic to tell someone, who would listen without judgment, how Facebook caused them to behave differently than they normally would and how their online interactions profoundly affected their lives. Some had relationship issues due to inappropriate flirting, jealousy or attention-seeking behavior. Others experienced the loss of friendships due to passive-aggressive posts, "unfriending," or communication misunderstandings. Others reported feeling overwhelmed by Facebook and tried taking Facebook vacations only to find themselves unable to break away because they were fully addicted to checking their News Feeds.

My goal in writing this book is to help readers understand how Facebook is affecting us personally and globally. Rather than providing a theoretical and data-filled account of social-media studies, this book provides real accounts from people who have struggled with their Facebook interactions, and teaches how we can start to make sense of the many personality and social changes developing as a result of Facebook encounters.

What follows is an attempt to address the general problems that arise from Facebook through case studies and examination.

I have collected stories from adults and teens, who have shared their personal struggles with Facebook. I have changed the names of participants and made some minor changes in detail to protect the anonymity of those who have participated, but all the case studies are firsthand accounts of Facebook users' experiences. Through these case studies, we understand why people interact differently online, how such interactions can lead to negative emotional outcomes, and what steps can be taken to find a new sense of balance. I'll examine how Facebook is affecting our sense of ourselves and of privacy and how we connect and seek approval in a way we never have before. I'll also discuss romantic relationships, how teens use and are impacted by social media, and the very real phenomenon of Facebook addictions. Last, you will learn about five types of emotional manipulators who find Facebook the perfect medium in which to victimize people.

The questions I asked my interview subjects over three years led to many more questions: Why are some people neglecting their families, friends, relationships, work, and education in order to spend more time logged on to Facebook? Why do we feel comfortable sharing personal information in a public forum when we wouldn't reveal it to a person standing right in front of us? Are we posting a link to express our opinions, or are we sharing it in order to impress others? Is the status update governing our behavior? Exactly how much influence does Facebook have on our emotions, relationships and lives?

Why are we so *Facehooked?*

chapter one

Connecting in the Digital Age

I'm a different person on Facebook. I think everyone is in a way. I know this sounds crazy, but I like when people really notice me and what I post—like people are seeing a new me; a better me. I share something on Facebook about five times a day. It's a part of life now isn't it? But sometimes I will admit—life on Facebook is more interesting than reality. I'd rather be responding to a comment on my wall than speaking to the boring person in front of me.

Before the Dawn

ONE OF MY CLIENTS, SAM, CALLED IN A SEVERE PANIC—
he couldn't tell me why, but repeated over and over that he felt
completely out of control and needed to see me right away. As a
clinical psychologist, I'm used to working with people in distress
and recognized immediately this man was overwhelmed with grief.
After I helped Sam calm down, he told me that his fiancé, Lisa, had
broken their engagement. This was certainly traumatic, but what
struck me about his story wasn't so much that she had ended the
engagement, but in how she had chosen to do so.

Sam had logged into Facebook that evening and discovered
that Lisa had changed her relationship status from "engaged" to
"in a relationship." More important, the photo next to the "in a
relationship with" was no longer his, but that of his best friend.
He called Lisa as soon as he saw her status update and learned she
and his friend had been an item for the past three months and that
they both thought it was time to let him know—by changing their
Facebook statuses. Sam was so overtaken with heartache I feared
he might harm himself. When his anguish escalated into suicidal
thoughts, he agreed that he might be safer in a hospital. I admitted

him to the Emergency Room and asked that he be placed on suicide watch. He was hospitalized for four days.

Another client, Megan, called me because something tragic happened to her family. Megan's husband, John, was scrolling through his Facebook News Feed when he came across a photograph posted by one of his cousins. The photo showed John's parents' totaled car—the front of the car was caved in and the windows were smashed. The caption underneath the photograph read, "Oh my God. My aunt and uncle died in a car crash!" John discovered his parents died in a car crash—on Facebook. He became hysterical. He called his family members in Texas only to find out that the Facebook post was true; his parents had perished in a car crash near their home. Ultimately the shock was so devastating that he was taken to a psychiatric hospital for treatment.

A few months back I conducted an intake evaluation on Ray, an eighteen-year-old with depression and severe pain related to a gunshot injury. Ray had been shot after fighting with his stepfather, who was upset that Ray was running up his cellphone bill because he was on Facebook all the time. This night, his stepfather screamed at him to get off Facebook because he'd been scrolling through his News Feed for two hours. The fight turned physical and Ray's stepfather grabbed a gun and shot Ray in the foot. Not only will Ray never walk normally again, he's been ostracized by his mother and other family members, who blame him for his stepfather's incarceration.

These are clearly extreme cases of the damage that Facebook can inflict. For Sam and John, Facebook was almost a weapon used against them: people they loved and trusted used Facebook to inflict pain upon them, without thinking about what effect their posts would have.

However, these are also examples of how Facebook is changing the way some of us express ourselves on social-media networks. Every day psychologists are hearing clients discuss issues pertaining to Facebook. Whether it has to do with a passive-aggressive comment that someone has made, poor boundaries, stalking behavior or social-media addiction, Facebook is a new dimension in our clients' therapeutic presentations.

Perhaps it's the instant access we all suddenly have to one another, or perhaps it's the distancing factor within Facebook that makes us believe that what we post will not affect other people as much as it would in real life. But there's something about Facebook that is causing many of us to lose sight of who we are, changing how we wish to represent ourselves online, and, in some cases, allowing common sense and judgment to go out the window.

The most significant problematic changes I've observed regarding Facebook self-identity, self-expression, and social interaction are: the amount of time we spend editing or enhancing our self-image, our increased need for public expression and decreased need for privacy, the emphasis we are placing on performing for others through our profiles, and choosing to connect with others through our Facebook avatars while neglecting real-life interactions.

VIRTUAL OR REALITY?

At the heart of the problem with Facebook is confusion about what is real and what is virtual. Without real-life social cues, we can never be certain whether what we see posted on Facebook actually happened or whether a person's profile reflects who they really are.

Taking this up an existential notch, we can ask ourselves if our expression of ourselves on Facebook actually gives off more truth about our core nature? Think of it this way—when a Facebook user posts multiple, well-tailored selfies on a daily basis, we tend to see through them, in that their online behavior demonstrates an aspect of their insecurity or obsessive need to appear a certain way to their Facebook audience. Alternatively, if someone posts photos of themselves where they appear casual or even "sloppy," we can assess that they are perhaps more carefree in nature. Either way, self-expression on Facebook reflects some basic changes in the way we are presenting ourselves to the world.

PERFORMING FOR THE MASSES

Facebook has encouraged us to be more "out there" more often. For the first time in modern existence, we are a culture of public personas, wanting to be seen, heard and followed like never before. We are systematically losing our previous innate need for anonymity and privacy and preferring to connect with others collectively versus individually.

Thanks to digital connection, group discussion has taken on a whole new meaning. In the 1980s we had group email followed by the invention of newsgroups. Such groups laid the foundation for current web boards that covered an array of subjects. Almost anyone could digitally weigh-in on practically any subject. But the real launch to interactive digital expression was the development of social media in the 2000s.

The Web opened up an entire new universe to us, and when social media came into existence our understanding of "group discussion" changed forever.

Since that time we've started changing in other ways. Gradually, people started paying more attention to building and nurturing their social-media relationships. Other changes involve our willingness to give up our independence by chaining ourselves to our smartphones and choosing to become quasi-present in front of others in real life. In today's world you can reach anyone, anytime and anywhere, and we feel a deep-seated obligation to respond to others in that same instant.

Before social media, we did not assume that we should respond to any kind of message immediately. Getting back to people took a backseat to whatever we were doing at the time. Now, as a culture, we've accepted that texting, video chatting, and instant messaging are to be respected, and that means that you stop whatever you are doing in real life, pick up your smartphone and respond to whatever push notification you are receiving. You may be sitting next to someone at dinner while responding to a Facebook comment. Although you are physically present, you are mentally engaged in your Facebook world.

Facebook has caused many to systematically avoid real-life interactions. Some would argue that when we were introduced to television, we experienced the same half-present mindset; however, the difference between watching television and social media is that you can watch television while conversing with those around you about what you are watching. The shows you watch on TV are not sending you personal messages, but the nature of Facebook interactions require more individual attention. After all, people are leaving you messages on your own wall so you feel more obligated to respond to your digital friends, or fans.

These significant social/cultural changes in communication give weight to the idea that Facebook, on some level, is changing our authentic selves, relationships and social behaviors. So how did we get here?

THE BIRTH OF SOCIAL MEDIA

In order to gain a full understanding of Facebook's other influences, we first have to take a look at how this all began. Technology has generated vast new forms of connectivity and self-expression. Once limited to face-to-face conversation, today's communication travels at light speed. Everyone can weigh-in on any subject in a matter of seconds. There are almost no limits to the extent of information we can access about our world and the people in it.

The Digital Age we live in is vastly different than anything we have ever known before, and when most of us tuned into the Internet Superhighway, we were immediately captivated. The Internet opened up a new world to us. Some people were able to predict the power of connectivity, while others were skeptical that the Internet would even make a dent in the way people interact. In 1995, Newsweek Magazine headlined an article: *The Internet? Bah! Hype Alert: Why Cyberspace Isn't, and Never Will Be, Nirvana...* Oops. Yet right around this time blogging began. Google was created a few years later.

What followed was an explosion of information and networking systems. In 2001, Wikipedia was created and Apple started selling iPods. Friendster, one of the first social-networking websites, was opened to the U.S. public and grew to three-million users in three months. MySpace followed in 2003, and the popular virtual world, Second Life, was launched. By then, there were more than three-billion webpages and Apple had introduced us

Timeline | *Events In Social Media History*

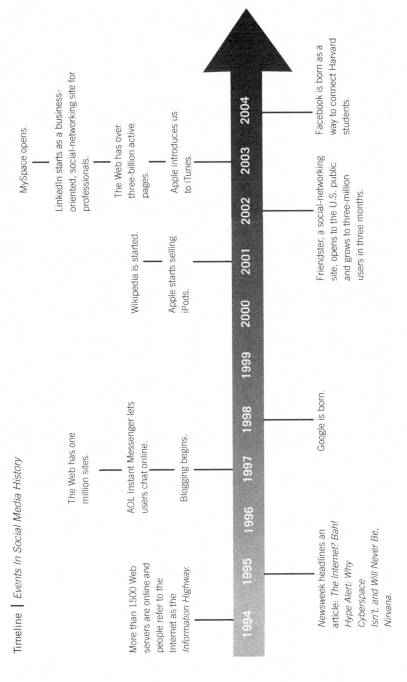

1994 More than 1500 Web servers are online and people refer to the Internet as the *Information Highway.*

1995 *Newsweek* headlines an article: *The Internet? Bah! Hype Alert: Why Cyberspace Isn't, and Will Never Be, Nirvana.*

1996 The Web has one million sites.

1997 AOL Instant Messenger lets users chat online.

Blogging begins.

MySpace opens.

1998 Google is born.

2001 Wikipedia is started.

Apple starts selling iPods.

LinkedIn starts as a business-oriented, social-networking site for professionals.

2002 Friendster, a social-networking site, opens to the U.S. public and grows to three-million users in three months.

2003 The Web has over three-billion active pages.

Apple introduces us to iTunes.

2004 Facebook is born as a way to connect Harvard students.

to iTunes. In 2004, Facebook was born, and five years later it was ranked (and is still currently ranked) as the most-used social network worldwide.

This social-media giant has expanded to include people from all ages. Facebook has evolved from simple college interactions and flirtations to the main source of connectivity for billions of people across the globe. Nowadays, pre-teens and senior citizens are discovering new ways to express themselves through their selfies and Facebook updates, while teenagers are migrating to other social networks. Today's Millennials are the masters of the social-media universe and unwittingly find themselves the main market for many corporations with an online presence.

Every major corporation understands that if they have any chance of striving and surviving, they'd better have a Facebook presence. Not only have corporations embraced Facebook, they count on it to tell them everything they want to know about their consumers. Big corporations have become the "Big Brother" of our time—observing and recording our every move, and we've allowed ourselves to become exhibitionists. We let them watch.

People have different comfort levels with their privacy settings. Some people try to restrict their Facebook pages to include only their closest family members and friends, but the truth is, a huge majority of Facebook users simply don't care who has access to their personal information or the intimate details about their relationships. On a collective cultural level, we have accepted that our deepest thoughts and feelings should be shared openly, and are subject to outside interpretation and judgment. This is who we are now. Whether we like to admit it or not, Facebook is now an important part of our existence.

Human nature causes the majority of us to resist change—

even positive change. Therefore it makes sense that whenever we are introduced to a new form of communication, some of us will accept this change, while others will express concern regarding how such advances affect our relationships. Both viewpoints reveal an understanding that social media has a profound effect on us. Eventually, we adjust to this new form of expression until it becomes second nature.

Face-to-face discussions, landline phone calls, newspapers and mail service have become replaced with smartphones, texting, photo sharing, instant messaging, gaming, and video chatting. And sharing news with each other within group settings has been replaced with Facebook updates, Twitter tweets, Instagram and Pinterest photos, and so on. The Digital Age has rapidly transformed the way we encounter one another.

THE HOOK

According to a 2012 Science Direct study, phantom vibration—a phenomenon where you think your phone is vibrating but it's not—has existed only since the launch of mobile phones. This "syndrome" is a sign of the digital intrusion in our lives. Today, it's so common that researchers have devoted studies to it. Research shows phantom vibration syndrome, or its other nicknames—hypovibochondria or ring-xiety—are a near-universal experience for people with smartphones. "Something in your brain is being triggered that's different than what was triggered just a few short years ago," says Dr. Larry Rosen, a research psychologist who studies how technology affects our minds.

Facebook and other social-media commentary trigger push notifications that are leading us to reach for our phones with the same speed that a child reaches for candy. This compulsive behavior is leading to obsession and other anxiety-related symptoms. People who constantly pick up their phones look like they have a compulsion, not much different from someone who repeatedly checks locks or washes their hands. When checking and rechecking our Facebook News Feeds becomes an addictive behavior, the only way to counter this effect is to give Facebook and our smartphones a rest. And just like Dr. Rosen, I am all for technological advancement, but Facebook users must stop spending so much time plugged in. When it comes to social media, we either love it or hate it, and even when we hate it, many of us can't seem to put it aside for long periods of time.

Facebook's driving force is that it allows us to exchange messages without being physically present. Facebook gives us almost unlimited expression and connection. On a certain level, the power behind social media is awe-inspiring. Aside from connectivity, Facebook allows us to give and receive information like never before. All of a sudden, we want to know everything about each other and we want to share more of ourselves with the world. This kind of connectivity and information sharing is certainly empowering, but it's also changing who we are, how we interact, and how we perceive our world. Sharing on Facebook is no longer just a bridge between our thoughts and the world; Facebook has become an alternative reality.

LOSING OUR GRIP WITH REALITY

Nowadays, many of us are splitting our time between Facebook and interacting in the real world, and the two realities are slowly merging. If we share something on Facebook we want to tell our friends about it, and when something entertaining happens to us in real life, we feel the need to share it on Facebook. This seems harmless enough until we begin to substitute one reality for another. Facebook has caused some of our behaviors to change. Aside from feeling a compulsion to check our News Feeds and keep posting photos of ourselves, we tend to be more provocative on Facebook. Some of us enjoy the new found courage we feel when expressing ourselves online in a way that we never would in real life. Some people even feel that their truest nature is best expressed on Facebook. In many ways we are creating a new self or a new identity, and this new self is also functioning as a new social understanding.

The ability to communicate with many different people across the globe at record speeds is changing our social interpretation of the messages we send and receive. Interacting with others compulsively on Facebook is changing the deep-seated and intuitive understanding we have about how to engage with others. Digital communication shelters us from certain human experiences. For example, thanks to our online interactions, we can now avoid the discomfort or anxiety we feel when confronting someone in real life, flirting or breaking up with a significant other.

What happens when we begin forming new assumptions about life, love and friendship based on our Facebook reality, or when our online interactions affect our real-life relationships? And what happens when the selves created through Facebook don't match the selves we present face-to-face, or even worse, when they contradict each other? In psychological terms we will experience *cognitive dissonance*, which is the anxiety we feel from us holding two conflicting ways of perceiving our world. Such a discrepancy between our perceptions and beliefs throws us off balance emotionally, and will inevitably lead us to experience identity confusion, relationship conflicts, changes in our judgment and, at an extreme level, even a psychotic break. When this happens, something must radically change in our perception in order to eliminate or reduce the confusion. We need to discover what kind of meaning we're placing on our digital relationships and then find balance between our former unwired selves and our new digital interactions.

We are still lacking a lot of research in terms of Facebook's psychological effects. And research can barely keep up with the pace of changes in social media. So Facebook's effect on us is a fluid one—just as Facebook evolves, our responses to these changes evolve along with it, but one thing is unmistakable: Social media is here to stay, and with over one-billion active users worldwide, Facebook has become an integral part of our landscape, especially for younger generations. It's vital to our identities, friendships, romantic partnerships, and family life.

It's crucial that we recognize how social media can influence us psychologically. It is my hope that this awareness can lead us to start making conscious decisions that will bring about a better balance between our online and offline realities. Facebook has brought some positive effects to people's social lives, to be sure. However, as a clinical psychologist, I was drawn to those cases where behaviors on Facebook were toxic and destructive, where people's lives were knocked so out of balance they ended up as clients in my office.

Before we can explore how Facebook is affecting the way we are relating to one another, we must take a look at how Facebook is affecting our self-identity. In real life, we can play different roles for different people, but on Facebook, many of us are actually taking on new roles with the ability to create, exaggerate and edit our "selves." In the next chapter we will see how Facebook sets the stage for us to perform for others, and in doing so, many of us are choosing to reinvent ourselves, choosing the characteristics of our created avatars, over acceptance of our true selves.

chapter two

Am I My Profile Pic?

I think I spend so much time on Facebook because my Facebook self is the person I always wanted to be: smarter, funnier, and more beautiful. On Facebook, I present the person I was born to be. I can change things about my life and I can change who I really am. And no one will ever have to know my mistakes or see my faults.

FACEBOOK AND OUR SENSE OF IDENTITY

WE HAVE FOUND OURSELVES IN A WORLD WHERE FACEBOOK and other social-media venues are changing the way some of us express ourselves online and in person. Every day psychologists are hearing clients discuss issues pertaining to Facebook. Whether it has to do with a passive-aggressive comment that someone has made, poor boundaries, stalking behavior or social-media addiction, Facebook adds a new layer to our clients' therapeutic presentations.

SELF-EXPRESSION OR SELF-EDITING?

While some people are not giving any thought to what they share online, many of us are. In fact, many of us spend a lot of time thinking about how to express or represent ourselves. How often have you started to write a Facebook post only to pause, reflect, backspace, delete and edit your original thought? In this new and ever-changing, social-media world it's hard to always know what is okay to post and what may be a Facebook faux pas.

When I first signed on to Facebook I shared an awkward post, *I have a cold*—mostly because truthfully I didn't know what I was supposed to post. This was a new world to me. I looked to my friends' posts for direction on Facebook etiquette. Initially I didn't like the idea of sharing my thoughts, feelings, or actions in such a public manner, but as time went on, I noticed my comfort level with my privacy changing. I found I wanted to express myself more openly than ever before. I attributed this to the fact that, on Facebook, someone is always watching, and as any social psychologist will tell you, when we are being watched, our behavior changes. When we have an audience, we are more prone to "perform"—very much like being on a stage of our own creation.

Suddenly it seems we are on display for the world to see and this changes things somehow. Facebook users have all encountered a situation where one of our Facebook friends shared a particular post and we wonder what they were thinking when they decided to share something so shocking. Why are we stunned by certain Facebook posts or selfies (photos that we take of ourselves—usually through our smartphones)? Perhaps it's because what we see on our News Feed in no way represents our friend, as we know them, in real life. And it's not just our friends, is it? Before Facebook we didn't seem so prone to focus on our self-image, so why now? Why have we suddenly become consumed with sharing more photographs of ourselves? Why are we expressing ourselves in a way that seems bolder, more "in your face," or downright wacky?

On one hand, what our friends post can be simply a reflection and update of their day. This is perfectly normal and enjoyable for many people. On the other hand, many people are feeling more and more compelled to share only the positive aspects of themselves while hiding what they perceive to be negative.

Bob, 41

New York City, New York

I'm handsome so I constantly update my profile photo with a new photo of myself. If others are in the photo I crop them out so that only my face appears. In my profile bio, I self-identify as a 'jet-setter,' often flying to Miami, LA, and London where I rub shoulders with politicians and famous people who are known for not having last names. I place myself in the 'super model' league with a penchant for high-end shopping. My profile contains a photo album of previous wall photos, many of which, like my profile pics, are solely of me. Those photos that don't showcase my good looks show things like the parking lot sign identifying my named parking space at the office or a BluntCard.com card that says, 'A nice thing I like to do is walk around the office so that everyone can enjoy how handsome I am.' One of my recent 'shares' on Facebook is an advertisement for a local beauty salon with instructions to watch for me in their commercial. My favorite quote is 'I drink to make other people more interesting.'

Like   Comment   Share

Nearly everything in Bob's profile is posted to solicit positive feedback from others. We've all come across someone like this, haven't we? Bob's photographs are always flattering, and in them he is almost always posed. His wall posts, articles or videos seem to always involve news stories about the high-profile social events he attends. It would seem that his profile is a channel for him to receive validation from his friends. What is striking about Bob's profile is the almost constant emphasis he places on trying to impress people. This need is just as obvious in his real-life interactions, but Facebook has become his vehicle for bringing him even more attention.

Naturally we come across many situations in real life where we like to put our best foot forward: first dates, job interviews, and meeting the in-laws for the first time. But for many, Facebook posting takes trying to impress people to another level. Every day on Facebook we see our friends posting about heading to the gym, doing charity work, caring for their elderly parents, returning to the gym, checking themselves into the newest and trendiest hot spots in the city, reading to their children before tucking them into bed, and then heading to the gym one last time.

I too fell under the Facebook posting spell. It's just too intoxicating to place yourself on display, or parade, and have your friends endorse almost everything that you do. This seems to be the main allure behind posting on Facebook—we can not only express ourselves like never before to so many people at the same time, but we can also decide what we share and how we share it in a manner that allows us to re-create ourselves. Don't like the weight you put on? No problem. Post a pic of you taken five years ago instead, and if that doesn't work, there's always Photoshop. On Facebook you have the ability to edit almost everything about yourself, and unless someone knows you in real life, you have a lot of leeway in terms of the life you create. This new possibility—the ability to re-create our identity—is a curious thing.

A while back I ran a marathon. I gladly posted photographs of me crossing the finish line in a blaze of (self-perceived) glory, but wasn't it all the other equally real and possibly less-flattering moments along the way that got me to the finish line? As I edited what I chose to share about the marathon I found myself focusing more on the end-result over the journey—something that I'm not normally prone to do. This was when I realized: Editing myself online has caused me to edit the way I perceive certain things about my life. This made sense. If I'm choosing to express myself only through certain photographs, then I'm also choosing to "see" myself through a specific lens.

The minute I caught myself tailoring what I shared on my wall, I questioned if Facebook somehow tapped into this deep-seated need of ours to hide our insecurities while embellishing or exaggerating our strengths, assets or accomplishments? Have you ever felt anxiety about what you've posted and wondered if you should delete it? Where does this anxiety come from?

In terms of the emotional effects that social media has on many of us, this is the question that currently challenges us psychologists. Why are we not okay expressing ourselves online as we do offline? Are we experiencing more pressure to appear funny, intelligent or entertaining? Are we beginning to lose a sense of self-acceptance, and if so what is triggering this self-doubt?

I believe that we are altering ourselves on Facebook because when we see our life displayed right in front of us—our profile appearing in our "face," so to speak—many of us feel the need to express ourselves through a better version of ourselves. Whenever we are under public scrutiny, we feel a need to self-edit and doubt our automatic self-expression.

Writers can create some amazing things on paper when they allow themselves to let go of the need to edit themselves during the creative process. When they can write solely to write, without the lingering burden of knowing someone will read their work, they find themselves amazed at their own creation. However, the minute the "self-editor" appears, they lose the pure expression of their vision. Similarly, the same applies to expression on Facebook and social media. We can easily lose sight of who we really are out of fear of how we appear to others. When self-doubt enters the picture, our self-editor goes into action, and we end up vastly misrepresented.

Life is hard enough. We don't need another reason to doubt ourselves, and yet many of us doubt what we post before we even post it. We're collectively engaging in this online sandbox where we create an imaginary world, where we spend time thinking of ways to embellish ourselves and become better than we believe ourselves to be in real life.

CONNECTING THROUGH A NEW PERSONA

Facebook is nothing more than an extension of childlike imagination expressing itself in an adult world. This is part of human nature; at best, simple, harmless fun. At worst, Facebook can be used to build an identity very different from who we are, can be used to lie, emotionally manipulate, begin false relationships, end marriages, and stalk and harass others (more on emotional manipulation later). Some people, of course, can develop a fake identity in real life, but somehow Facebook encourages a façade, urging us to amplify our imaginary selves while denying our true personalities.

While there are many benefits to self-expression and public connection through social media, it's troubling how social-media interaction is affecting what psychologists refer to as our *self-concept* or sense of self. In real life, we typically form our self-

concept through our interactions with others. We learn who we
are and what we believe in through our social interactions and by
comparing our understanding of the world against the messages we
receive from others. Consider this: What if the messages we receive
about ourselves are transmitted instantly, without limits and are
constantly reversing themselves? What then? Eventually we'd
become confused and even anxious about the messages, and on an
extreme level we wouldn't necessarily know which messages were
real and which were not.

What is schizophrenia but the inability to make the distinction
between reality and hallucination? Every day on Facebook we
expose ourselves to mixed messages received from multiple
sources at the same time. It's no wonder that more and more people
report experiencing anxiety and depression due to their Facebook
interactions. The bombardment of messages we receive is a lot to
take in sometimes. A woman posted this commentary on experien-
ceproject.com about stalking Facebook users' photographs:

*Yes, I have and I feel bad about it because it causes guilt,
envy, jealousy, and discontent with your own life. When I am
bored and just browsing around on Facebook, I sometimes
just look at people's profiles because I get curious about
people, and especially those I don't know. Then it seems
like these other people have better jobs, nicer homes, better
this and that, etc., and it does nothing but make me feel
horrible and guilty all at the same time. One week I was
even crying because I felt my husband could never be as
wonderful as this other woman's husband, and what was
stupid about that was, I had never even MET these people;
in fact, they lived far away. Since then I have stopped*

looking at Facebook profiles altogether, and only look at my News Feed from friends I know, which is enjoyable.

Even those of us who don't exaggerate who we are online, edit ourselves in some way. And the more we edit ourselves to reflect who we *think* we should be, the more we lose sight of our true persona. Remember how you felt in high school? If we weren't a part of the super cliques, many of us spend some part of our adulthood trying to forget our teen years. Generally teens determine their self-worth based on the opinions of their peers. Facebook endorsement and validation is not so different. Many people try to post photos,

Helena, 33

Savannah, Georgia

My best friend loves posting all our pictures on Facebook. For the most part, I don't mind having my picture taken, it's just that I don't want all of my photos shared on Facebook. In some pictures I look great, in others, not so great. I get nervous when I get tagged in unflattering pics, 'What if I look horrible?' It's like I have no control over what people see. I asked my friend not to post any picture of me without my approval and she agreed, although she thinks I'm being silly. I feel foolish for even asking my friend for this favor, but I can't help myself. When I see bad photos, I don't want people to think I really look like that, to think that's really me, even if it IS the real me.

⌃ Like　　　🗨 Comment　　　➦ Share

videos, links or comments that will get them "likes," and some are consumed with tailoring their self-image on Facebook to garner that approval. The emotional effect of Facebook persona editing can result in low self-esteem, depression, and an overreliance on what other people think to determine how we should act, what we should think, and who we should be.

While we certainly can create an alternative persona in real life, the difference with Facebook is that we're always "on." We may be able to change into jeans and a comfy T-shirt and lounge around the house without caring how we look. But on Facebook, we always care. Even when not logged on, our profiles are still up for others to see. The online stage performance continues even when we think we've "closed the curtain."

Connection or Control?

The other day I was on the train heading to my office and I noticed that almost everyone around me was avoiding making eye contact. This is certainly not unusual for Chicago, but what caught my attention is where everyone's eyes were focused. Almost everyone was staring at their smartphones and scrolling down the feed of different social-media sites, including Facebook. It seemed a bit ironic. People were connecting to other people, just not to anyone around them.

What makes us connect to other people anyway? What is it exactly that pushes us to walk up to a complete stranger and start talking? Maybe it's the way they talk or the way they laugh. We tend to form initial impressions of people within seconds. In particular, we form our first impressions based on people's facial features.

Sizing people up based on their appearance is human nature.

We do this every day in different situations. On Facebook it's not so different, except for one thing: On Facebook we initially decide whom to interact with based on the *selected* images they've presented for us. We feel a sense of control when we can choose which parts of ourselves to present to the world. But do we really have control when we worry about how we will be publicly perceived? We fret over whether to share the pictures of ourselves in all our high-school glory, with 80s hair or 90s grunge fashion or a recent selfie; to let people know that we really like watching bad sci-fi movies or pretend that we prefer intelligent documentaries; share our dog's misbehavior or just post the photos where she's posed under the Christmas tree? Which page is really more authentic, and which is worthy enough of trying to hook someone into our own Facebook world?

Developmentally, most of us learn through our life experiences to not care so much about what other people think, but Facebook has somehow taken us back to a point in time where what we publicly post seems to take precedent over who we are in real life. Instead of focusing on shaping who we are or creating who we wish to become, many of us focus on creating and crafting our Facebook avatars.

People's false sense of control is expressed through their need to re-image themselves: If I can predict your reaction to my photo, then I have control over how I appear. Obtaining this type of control in real life is extremely difficult if not impossible. Therefore, many choose to try to gain this sense of control through a façade or non-reality, a place where you don't have to face the truth about yourself or others. This is the "hookable" aspect of Facebook. Many of us know we are only seeing a certain part of our Facebook friends' lives, but we don't care. We are beginning to live for positive online affirmations and endorsements while deleting the comments from our profile that we find less-than-affirming.

The Facebook Avatar—The New Reality Star

Interacting and engaging through our Facebook avatars is somehow preferable than interacting in reality. With a simple click of a button I can remove your negative feedback as if it never happened, or I can selectively post the photographs about my life that make me feel a certain way. The minute we choose to censor or remove certain truths about ourselves—whether online or off—we are consciously choosing to put blinders on to the parts of ourselves that may require our attention.

Gabriela, 27

Omaha, Nebraska

I noticed my sister Teresa becoming more and more active on Facebook. In particular, she started posting many photos of herself with her husband and their children. At first, I didn't notice anything special about Teresa's postings, but I became curious when she kept adding the same kinds of photos. And then Teresa started to post various love poems dedicated to him, saying what an 'amazing husband' she has and tagging him in many of their wedding photographs. One day I told her how pleased I was that Teresa and her husband were so happy together. Teresa broke down in tears and told me that her husband recently asked her for a separation; he was seeing someone else. Teresa posted all of their happy photos on her Facebook wall in order to hide from the reality of her marriage.

ఠ Like   🗨 Comment   ↱ Share

One out of every four profiles contains false personal information. With a quick edit, one can become a successful professional, a world traveler, an all-knowing guru, a fantastic lover, a muscle-bound Adonis, or a hot supermodel. Without the benefit of having met the person in real life, we could be easily fooled into believing what we see in their profile. What happens when we get hooked to a false identity? What are we to do when our façades become the main reference points for friends, prospective bosses, or romantic relationships?

Facebook presents us with a layer of perception that takes us further from the reality of the situation. If the beautiful woman sitting at the next table with perfectly coiffed hair, stylish clothes and beautiful jewelry makes me feel self-conscious about my own appearance, then I'm comparing myself to her as she actually is. If, on the other hand, viewing someone's profile pic on Facebook makes me feel self-conscious about how I look, I'm comparing my actual-self to her representation online. It is not a real, one-to-one comparison—I'm quite sure that I could create a profile pic just as stylish as hers. The emotional effect in each scenario is just as real—my self-esteem takes a hit.

A recent study found that many people feel envious and resentful of others after seeing pictures of weddings, vacations, and other happy events on Facebook, and about one-third of all users feel worse about themselves after surfing the site. Those who unsuccessfully compare their lives to others' based on what they see on Facebook find their lives downright boring and less successful. Self-esteem weakens, and when this weakening becomes persistent, the end result is a pattern of self-defeating behaviors. For example, when someone stops caring for themselves, they will end up spiraling further down the negative slope of defeating thoughts. First, you believe your life isn't very interesting. Then you start comparing how successful you are to your friends. Then you begin to question how happy you are in your relationship. Negative thoughts lead to negative feelings, negative feelings lead to poor decisions, and poor decisions lead to further unhappiness. The interconnectedness within Facebook is subtle yet effective. On Facebook, we are all connected and have an immense effect on one another.

Social comparison is nothing new; we've always tried to "keep up with the Joneses," and Facebook simply provides another way to do so. What Facebook doesn't give us is a complete picture of someone else's life—and why should it? While some of us are willing to showcase our failures and embarrassments, most people appear to use Facebook as a showcase in which they display their successes and trophies for others to see and appreciate.

On Facebook, the photo taken of your friends out drinking and dancing on New Year's Eve wasn't followed up with their hangover photo. Your best girlfriend's profile photo where she's wearing the latest Jimmy Choos was actually of her trying them on in the store—she never bought them. Your friend posing for a picture at the club surrounded by hot women was taken in order to try to make his girlfriend jealous.

You cannot make assumptions about someone's life based on posts they share about their charity work, their straight "A" student, or their vacations. Think about it—generally, are the majority of your friends' posts related to their insecurities, jealous tendencies, control issues or poor dating record? Believing that people are living an optimal life based on what they post is dangerous because sooner or later you will begin to compare your life to theirs and when you do you are functioning out of a false reality.

WHERE DO WE GO FROM HERE?

A Facebook profile is not supposed to actually come to life, but many people begin to shape their real lives around how their profile describes them. Instead of living, some of us are just "performing." We act in ways that encourage responses from others—perfect performance—but no one is perfect in real life. What would it be like to be involved in the real world the way you are on Facebook? Instead of portraying yourself in a band, how about actually joining a band? Instead of spending so much time focusing on presenting a "perfect" image, focus instead on what actually is—the ups and downs that are real life. Try posting silly photos of yourself, and laugh with your friends when they laugh. If you do something embarrassing, post it! Share it with others, enjoy the laughter together, and stop worrying about what others will think.

Daniel, 35

Melbourne, Australia

I posted a photo of myself playing a guitar with a caption, 'Me at band practice.' Shortly afterward, a friend busted me by commenting that I wasn't in a band and that I was just posing with a guitar. I quickly deleted his comment. A few months later, I tried to impress an old girlfriend by reposting the same picture on my timeline so that it appeared current. I added the same caption, 'Me at band practice,' prompting friends who knew better to once again make fun of me.

      👍 Like   💬 Comment   ➦ Share

We've all heard that we should "be true to yourself." But what exactly does this mean? Be sincere, present and authentic. With Facebook, this is an incredible challenge: it offers you a filter for what you want to say and how you say it, how you look and how you want to appear to the world. There's planning involved in the editing process, so the end result is a carefully edited version of yourself in your profile. The more you self-edit, the less you end up valuing your actual self, and the worst mistake you can make is to place more emphasis on the opinions of other people instead of your own inner voice.

What do you like most about yourself? Write it down. If you're going to represent yourself in any way, doesn't it make sense to be truthful about yourself and your current lifestyle? Who cares what you did or how you looked five years ago? What are you up to now? What are your plans for the future? Spell them out, even if they sound crazy or irrational. What makes you, you? Embrace this with all of your might in every situation.

Take some time to look over your profile as it is now. Seriously, put this book (or eBook) down and go to Facebook and click on your profile. Ask yourself—and your friends—if it truly represents who you are. Your true friends will give you an honest appraisal and appreciate who you are. Free yourself from self-imposed expectations and begin to celebrate a more genuine you.

chapter three

If It's On My Wall, It's Private, Right?

I need people to see me and to comment on what I do. There's no point otherwise. Why would I want to do something if no one will notice me doing it?

Privacy and Facebook Settings

WE LOVE TALKING ABOUT OURSELVES. IN FACT, ACCORDING to *Time Magazine*, researchers from Harvard University Social Cognitive and Affective Neuroscience Lab discovered that talking about ourselves triggers increased activity in the same parts of our brains as occurs when we're eating or having sex. Think about the last time you were having a conversation and someone said something that caused you to think about your own experience. Did it trigger emotions that caused you to want to react? Did you find yourself patiently (or not so patiently) waiting for them to finish talking so you could then share your story with them?

We seek out audiences because we tend to understand other people's situations, and even our own behavior through conversations. This is the main reason why people come to therapy: *Please listen to my story and tell me what you think.* If someone agrees with us, we're happy, knowing that we made a "correct" decision. When they disagree, we experience uneasiness.

Another reason we enjoy talking about ourselves is because we love to include our friends in an entertaining story. If we find something funny, we believe our friends will too. Every time we share something with our friends, there is a feeling of connectedness and understanding. On another level, our activities don't seem as "real" until we share them, and sharing them requires an audience. Our need to self-disclose is very much like the empty forest riddle: If you tweet in an empty forest and no one is there to read it, does your message have meaning?

These needs to expose aspects of ourselves aren't so surprising. All you have to do is attend a cocktail party. In almost every corner you will hear conversations about a person's life, their family, their jobs, how many networking contacts they have, the parties they attended, and how crazy they got last Friday night. People get excited when given the opportunity to share their story. And along comes the Facebook "gathering" to fulfill this need we all have.

Every minute on Facebook walls, we see people post about their latest experiences. We read about our friend's new haircut, their cat's health, their all-star parking spot, their hot new girlfriend, the horrible blind date, and their child's debut in the school play. At other times we read about our friend's more personal thoughts on certain subjects; thoughts perhaps more appropriately confined to a diary: *My family members are all crazy. Why can't I find a good man? My neighbor needs some serious therapy. His wedding ring is off...guess I know what that means.* The emotional consequences of sharing such personal information publicly can be severe—and yet many of us keep on sharing.

Metaphorically, the keyboard has become a gateway between our thoughts and the public, while the computer screen has morphed into an emotional shield. Perceived safety and unlimited public access encourages us to openly broadcast our most private thoughts and feelings. And while we may feel empowered by our ability to express ourselves like never before, sharing our thoughts without filter can leave us feeling overly exposed. When we find that our self-expression gets us (or others) into trouble, we may be pushing the limits of what's considered healthy communication. Ask yourself this: Is your emotional expression based on a need to elevate others or a need to diminish others to elevate yourself? Are you possibly hurting yourself in the process? Here's a thought: What if oversharing emotional expression on Facebook actually promotes problems instead of alleviating them. What happens then?

Maura, 36

Chicago, Illinois

I noticed my husband, Charles, was becoming increasingly distant. He started spending a lot of time on Facebook. Whenever I'd ask him what he was doing or with whom he was talking, he would quickly close his laptop. Each attempt I made to involve myself more in his life was rejected. I avoided what I already knew: Charles was having an affair and it was with someone that he frequently interacted with on Facebook. I began noticing a woman adding flirtatious comments on his posts. When I checked this woman's Facebook wall, I saw comments alluding to her feelings for my husband. I started sharing my own feelings on my wall. Way deep inside, I knew I was wrong to do this, especially since Charles and I shared many mutual friends, but somehow I couldn't help myself. If this woman was going public, so could I! After all, I had feelings too.

🖒 Like   🗨 Comment   ➦ Share

As you would guess, there's a wide range of what some people are willing to share on social media. Some people avoid this feature entirely; some simply update their status with their latest news, while others leave nothing to the imagination. Whatever thoughts they are having, whatever feelings they are experiencing, whatever situation they are going through, friends on Facebook will become aware of it. No matter how much or how little is publicly shared on one's profile; every post will inevitably be examined and judged. But somehow, we don't tend to mind the judgment as long as we are heard, and there are plenty of us who want to read all about it.

Perhaps a form of stalking is becoming our new favorite pastime. Reality shows certainly have given us plenty of opportunities to spy into the lives of others. How many of us have become addicted to at least one? I've definitely needed my "fix" of a few. I've often wondered which need is stronger: the need to express oneself or the need to watch the expression of others? Facebook fulfills both needs at once, and the current technology toys (tablets, laptops, and smartphones) may be changing the game further, letting us quickly and constantly gain access into a theatre-like world where we can become both the performer and the audience.

Feedback Please?

The artist often strives for wider and wider audiences to interpret his work. The politician hopes to spread his message to as many people as possible. The businesswoman networks with organizations across the country. Why is having a vast audience important? Most people don't use expression simply for the sake of hearing their own voices. What is the drive that makes us so curious what others think? Don't we all need reactions? Don't we all need to know that if we push the envelope that you will indeed be shocked?

I don't believe that the immense level of self-disclosure on social-media sites is solely about satisfying a basic, social-connection need. Rather, we may be broadcasting our thoughts and beliefs over a larger spectrum of people in the hope that someone (or many someones) will respond to us directly. After all, the holy grail of Twitter is the direct response message or the public shout out using @ in our own tweets. Will you respond to me on a group level, or will I make enough of an impact that you'll want to make direct contact? How much influence can I have over your life? And, in return, how much will you share with me?

As we have advanced in technology, we've grown accustomed to more of everything: more speed, more information and even more access into someone's personal life. We like having easy access to just about everything we can think of so we've certainly gained a lot in terms of speed and instant gratification, but how much have we lost? We've moved from critically examining how to express ourselves to impulsively wanting our thoughts acknowledged. We've gone from respecting others' privacy to expecting more exposure. And this communication shift undoubtedly causes us to interact in a public forum more often than we did before, and to less frequently maintain some private aspects of our lives.

The more we use social media and gain greater familiarity with what others post, the more we tend to allow ourselves to explore our own emotions, needs, and wants. It is through our self-expression that we get to know ourselves, and the more we share the more we learn—even if that means we learn about the darker sides of our personality, the parts of ourselves that we would normally prefer to remain hidden.

Sharing our private lives on Facebook may be risky on some level, but it is tantalizing at the same time. Sometimes the anticipation of others' comments and likes can be just as exciting as the comments themselves. We can't wait to know what other people think of us, because how else would we know ourselves? The philosopher Immanuel Kant taught us that our self-awareness is a direct consequence of the existence of things outside of us: We get to know ourselves only through our interactions, and therefore we are dependent on the feedback of others.

Is It Worth It?

The more we post, the more our decisions, lifestyles and beliefs face scrutiny. How many of us can stand this scrutiny or even want to? Some people seem to like the attention. In fact, some people even seek bold responses by posting over-the-top commentary and remarks intended to spark controversy. On the other hand, many of us don't like to be judged and may not be prepared to hear the truth about ourselves. On some level this seems awfully hypocritical. If we willingly put our opinions out there, why would we expect our friends to remain mute with theirs? Isn't the point of social media to publicly interact with each other? Doesn't social media interaction expose us to different ways of thinking?

Now, more than ever, it seems that losing our privacy is less important than what we believe we gain from self-disclosure: closer alliances with like-minded individuals. Many of us prefer to share our stories with those who can understand us. Psychologically, many of us feel a strong need to engage with people who agree with us. We tend to only visit websites that agree with our political opinions, and we usually spend time around people who hold similar views and tastes. We also tend to avoid individuals, groups, and news sources that challenge our views.

There's comfort in knowing that friends are entertained by status updates, that they appreciate our views on society and politics and that they "like" and endorse our way of thinking. When we feel understood we feel appreciated, validated, even adored, but this adoration comes at an emotional cost.

Aside from making our profiles public, Facebook made changes that allow others more access to our posts. Some tracking apps allow users to make search inquiries into almost any topic, making it easier to search for lovers outside of your Facebook friends. Facebook added facial recognition to the photo-tagging feature, making it easier to spot and tag people based on their profile pics, and they introduced embeddable posts in August of 2013, allowing news organizations and blogs to include public status updates, videos and photos in their own stories.

Let's face it, when it comes to social media, privacy has never been a social norm. In fact, Consumer Reports Internet Privacy and Security Survey reported that close to 13 million Facebook users have never even touched their privacy settings. This survey found that among one-billion active monthly users, 28 percent of them are sharing all of their wall posts with audiences greater than just their friends—whether they are aware of it or not.

Before Facebook and other social media, a person kept most experiences private, choosing what they wanted to share with their friends or family. We used to live in a world with two separate types of acts: private and public acts. Our understanding of privacy was simpler. What we did in our own bedrooms was our own business; what we did outside our homes was open for others to see. How does this complex new web get in the way of your privacy?

One of the things that a free-for-all privacy standard does is to give us a feeling—false in many ways—of being in more control when we know the "truth" about someone. Some people can't take things at face value and need evidence at all turns. When someone makes a statement about something, another may feel the need to

Paula, 55

New York City, New York

I was looking for a life partner, so I decided to use Facebook to meet men. I met someone I really liked. I enjoyed the flirtation between us and fell hard for this guy. I felt enamored with the prospect of romance. This man told me that he was in financial trouble and began telling me convoluted stories. First he said he lived in Italy and then said he was an engineer working in South America, but that his job had fallen through. Before long he started asking for money. He kept asking and it made me angry. I caught on to the scam but decided to play along to mess with him and get back at him. I was certain I was being played when he asked me to send him $30,000 in cash to an address in the Netherlands. Before I decided to block him, I sent out a message to his other friends warning them. Many of them shared similar stories of his attempts to romance them and then ask for money. I realized that I became a prospective scam victim when I made my profile public. Somehow I thought what I shared on Facebook would be safe and not used against me.

ᗷ Like   💬 Comment   ↪ Share make sure it's true by "Googling it" immediately—even right in front of them. Forget social decorum; having instant mobile access to Wikipedia, Google, and Facebook has somehow made it socially acceptable to fact-check, even if it makes you look like a jerk.

Were we better off with a simpler understanding of privacy? I believe we were. We enjoyed conversations simply for the sake of being in the company of interesting people. We learned about someone from what they said to us in person, not from what they posted on Facebook. Gaining information about a person from a public domain (versus hearing it directly from a person's mouth) may lead you to believe that what you see and read on social media is in fact true.

What should be completely private in the social-media age? Do we give up what we once held sacred, or do we try to reach for greater levels of privacy in the midst of a not-so-private way of communicating? Who is seeing our information anyway? Should we care if corporations and retailers are monitoring personal data?

So far, it seems that we have three main privacy "leaks"—what we see, what our friends see, and what Facebook sees. That last point is the most risky for those of us wanting more control over what we share. One of the reasons why Facebook's privacy settings keep changing is because marketers and advertisers all want access to the consumer Holy Grail: our demographics, our ages, how many children we have, where we live, what topics we search for, what colors we like, and, most important, where we shop.

Privacy on Facebook is a double-edged sword; the more we share, the more we want to protect. Yet, the more time we spend on social-media sites, the more we become accustomed to sharing more of ourselves. We all know this, but few of us stop sharing personal information and even fewer get off of Facebook. We judge people if they do. Think about it—how would you feel if someone you interacted with regularly suddenly got off Facebook entirely? The social pressure keeps us going, yet many of us do not realize (or care) how much we are becoming involved with our online identity.

Diego, 28

Mexico City, Mexico

My cousin Tony and I decided to start a business together. Times were tough and we had both lost our jobs. We trusted each other, hung out together since we were kids, and were practically brothers, so it made sense for us to join forces and begin a new sales career. Tony was the guy with the connections; I was the accounting guy. Everything was going smoothly until I noticed some irregularities in the accounts. Some money was missing; actually, too much money was missing. I talked to Tony about it but he couldn't explain what happened. I wanted to believe in Tony but it was hard. Then Tony posted on Facebook about his new suits and his new car. I knew he was stealing money from the company, and now he was bragging about it on Facebook! I had had enough so I posted on Tony's wall, 'Must be nice having such a sweet ride with all the money you stole from me.' Then all our relatives weighed-in on our situation. I found out later on that Tony wasn't stealing, but that Facebook post had already split our family in half. I felt bad about my post, but I just wish my family had stayed out of it.

    👍 Like   💬 Comment   ➡ Share

When did it become acceptable to publicly post our grievances? What does this do for us? How do we expect others to react to what we post? On Facebook we not only welcome self-disclosure, but we've grown to expect it. Some of us grumble at those who rarely post updates on their walls or who never comment on ours. But isn't the greatest faux paus when someone crosses the Facebook friendship line? As with offline friendships, a common unvoiced assumption for Facebook friendship etiquette is this: A person will remain my Facebook friend as long as they can be trusted with the information I share. If you tag or share a photo of mine without my permission, I may unfriend or block you. As long as you respect my "privacy" we can continue our social-media relationship, but considering the public nature of social-media sites, this assumption of privacy is a strong one to make.

An Uncontrollable Force

If we willingly choose to share a large amount of information about ourselves on Facebook, and other social media, how much information can we reasonably protect? Even with the tightest privacy settings it is unrealistic to expect that everything that we post will be secure. This false sense of security can influence people's behavior on Facebook. Can you think of some life event that you might not want to openly share? Ponder this: What if there were no privacy settings on social media? Would it change how much you share? I bet it would. Right now I'm imagining my mother being exposed to my funny, yet highly inappropriate, posts and wondering where her daughter went wrong. This is something to keep in mind, as outside lenses are allowed to zoom in on our privacy.

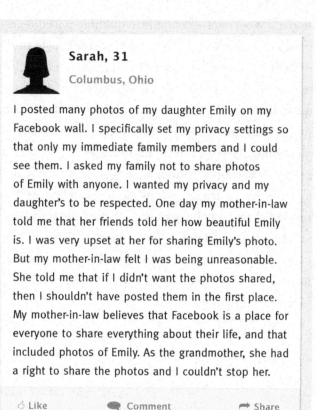

Sarah, 31

Columbus, Ohio

I posted many photos of my daughter Emily on my Facebook wall. I specifically set my privacy settings so that only my immediate family members and I could see them. I asked my family not to share photos of Emily with anyone. I wanted my privacy and my daughter's to be respected. One day my mother-in-law told me that her friends told her how beautiful Emily is. I was very upset at her for sharing Emily's photo. But my mother-in-law felt I was being unreasonable. She told me that if I didn't want the photos shared, then I shouldn't have posted them in the first place. My mother-in-law believes that Facebook is a place for everyone to share everything about their life, and that included photos of Emily. As the grandmother, she had a right to share the photos and I couldn't stop her.

🖒 Like   💬 Comment   ➦ Share

Big Brother

There are so many factors that play into what we can't control online. On a more global level we are discovering that our "private" information is not so private anymore. Edward Snowden, a former contractor of the U.S. National Security Agency (NSA) and a former employee of the Central Intelligence Agency (CIA) leaked top-secret details of U.S. and British governments. His leaks have been the subject of great controversy. Some people consider him a hero and whistleblower, while others describe him as a traitor. Similarly, Julian Assange, the editor-in-chief and founder of WikiLeaks, said this about social media, "Facebook, Google, and Yahoo are allowing the CIA to access user data via a specially designed interface. Facebook is the most appalling spying machine that has ever been invented." Through their actions and statements, these two individuals brought global awareness of just how easily outside sources can gain access to our private information.

Your Facebook profile may be filled with photos of your children, your thoughts about your in-laws, your plans to overthrow the government, and perhaps even a whole lot of evidence on the chapters of your life that you would, say, prefer to be "unpublished." It can be searched by your partner, stalked by your ex and probed by marketers, future colleges and employers. Facebook's new search tools even allow strangers to follow your next move. Your photos get tagged and people know where you are because of location check-ins. Nothing is sacred as long as you post it and our sense of control over what we post is an illusion. Think about the last time you posted something on Facebook, Instagram, or Twitter and deleted it later the same day just to realize that your post was already forwarded to those you hoped wouldn't see it. Perhaps you tried to alter your privacy settings. If you're like me, you probably went through what seemed like 27 steps to make sure what you post would only be seen by your Facebook friends. Then a few months later you realized that Facebook decided to change their privacy settings—again.

In a lot of ways, Facebook itself controls your privacy. This lack of control often leads to unfamiliar anxiety. Facebook maintains it is up to every user to decide how much others see and this is true, to some extent. You can tweak who has access to certain photos, but you cannot stop your friends from sharing them and talking about them behind your back. You may be able to limit who can see certain posts but you cannot stop someone from taking a photograph of your Facebook timeline and sharing it on their mobile devices. Additionally, with Facebook's new search features, there's simply no way that you can completely prohibit who can see what anymore. Everything you post nowadays is even easier to find. Realistically, the most that you can do is to limit your photos, edit what you share and decide just how much you really want to share on a public, social-media site.

All of this is not to say that I'm suggesting that we get off Facebook for lack of privacy. What I am suggesting is that perhaps we need to take a closer look at what we believe is private on social media and realize that we have less control over our privacy than we formerly thought.

BANG WITH FRIENDS

Branded "evilest app ever" and launched in January of 2013, Down, formerly known as "Bang With Friends," is a Facebook app you can use to see if your Facebook friends are willing to have sex with you. It allows a Facebook user to discover who from their friend list is open for "banging." Once a person responds to the choice made by one user, the site sends an approval and then the rest of the decision lies with the sender. Aside from the obvious lack in coyness from any Facebook user signing up for this app, it lets another user easily track your interests, activities and likes. On the other hand, this app can be a great way to hook up with people who are mutually interested. After all, what is the harm in using an app that makes finding a good romantic connection with others easier? The pros and cons of this app, and other hook-up apps that may compromise our privacy, depends on what we see as more important: sex or privacy.

Facebook overlooks any privacy settings for Down. So if you're cheating on your partner, your Facebook history can be tracked by other users. Anyone can trace which of their friends is using this app and what they are doing. On regular Facebook privacy settings, you can customize what is posted and who tags you in photos, but in Down your privacy settings do not apply. This app has obvious appeal for those searching for a sexual encounter or those seeking an opportunity to be wanted or desired, but what is shocking is the willingness of some users to disclose almost their entire Facebook history for the opportunity to "bang." Just how many concessions are we willing to make in order to use certain apps? Down shows that, with the right motivation, some people are all-too-willing to compromise their online privacy. For some the allure to connect socially, or sexually, is much more appealing than any other reason they would use Facebook.

CAN WE BE PRIVATE AND SOCIAL?

Being social and being private don't have to be mutually exclusive. When we are out with people in the bar scene we see everyone multitasking, speaking with people in person (and with others on their phones). The mobile access of Facebook has created a new contender while interacting with others in public. Never before has it been so easy to ignore the person sitting next to you on a train. In coffee shops we often see people hiding behind their laptops. Such interactions can be perceived to be private while being conducted in public. When it comes to Facebook, however, because our conversations are made in a public forum, our conversations themselves change from direct interactions to something on an open stage where people are no longer just active participants but also spectators.

What is troubling isn't so much the blurred lines between what is public and what is private, but the shifting understanding of what is acceptable to share publicly. As a global society, predisposed to limitless online communication received at vast speeds, we've neglected the preservation of what should be sacred and privileged information: intimate moments within our relationships, family secrets, workplace policies, and conversations made in confidence with those we trust. I believe that many of us have gotten a little too sloppy about guarding our privacy and have stopped making conscious decisions on what we choose to make public. We've lost some of our ability to truly step away from our online selves.

Facebook has changed our sense of privacy by altering our perceptions and actions in four big ways:

The illusion of control: When we believe that only we can control what we post, we are more likely to over share. Many people forget that when others have access to our photos or statuses they can choose to do whatever they want with them, including tagging, copying, sharing and reposting. It is usually only after someone has forwarded our information that we begin to feel the pinch of frustration that comes with others not respecting our privacy.

Detachment and distance: The sense of detachment we have around profiles and Facebook friends fosters in us an idea that the regular rules of society don't apply to Facebook. We believe that our posts don't carry as much weight in the real world as words spoken aloud. This disconnection encourages new behaviors stemming from the added sense of courage people feel when they don't have to face someone in person. Trust me folks, after a while, we all learn that the weight in both spaces can be quite heavy.

No Facebook "police": When you first sign on to Facebook there is an assumption that you should post at your own risk. No one hands you a rulebook on what or what not to share. There aren't clearly visible enforcers or authority figures, so people feel freer online than they do in real life. Yet, there can be serious repercussions to what we post. If we post anything that even hints at a threat to anyone, the police may well be knocking on our door before we know it.

Lack of immediate consequence to our behaviors: In face-to-face interactions, we immediately see or hear people's reactions to what we've said or done. But online, that's missing. On Facebook, there is often a gap in time between posts and responses, and some choose to just ignore a post they don't agree with rather than responding directly. For these reasons, the responses (or lack of them) don't appear to carry the same emotional effect as they do in real life.

WHERE DO WE GO FROM HERE?

If you are concerned about posts or photos that might embarrass you, comb through your timeline and get rid of them. So far, it seems that the only sure way to safeguard that a post or photo is not exposed is to "unlike" or "delete" it. You can always save funny, yet questionable photos in a separate computer folder. Every few months, look for profanity, references to alcohol, drugs and other indiscretions that you would rather not have a law school admissions officer or potential employer see, for example. Get comfortable and set aside an afternoon—this process is going to take quite a while, but it may be worth it, especially if you're in the process of finding work, applying to colleges, or preferring that your boss doesn't know about what you did last weekend.

Even if you set the highest privacy setting, it is naïve to assume that everything you post will remain private. Go to "who can see my stuff" on your Facebook page. Click on "see more settings." The default Facebook settings are set so that search engines can easily link to your timeline. Unless you want to broadcast your every move online, it may be best not to use the default. To a certain extent, by micromanaging your friend subgroups, you can determine who can view your photos, "likes" and status updates. Categorizing your friends on Facebook can take an awfully long time, but the end result is that you are putting yourself in a position of a *little* more control. If you don't know how to set stricter privacy settings, consult with Facebook friends who can help you set these up.

What is "safe" to post on Facebook? Every person's comfort level with self-disclosure is different. There are no clear and absolute guidelines for what should be made public, but my rule of thumb is this: If you're comfortable announcing the information shared in a post the same way as if you were announcing it through a bullhorn at your community grocery store, then go ahead and post it. If you are feeling reluctant in any way about posting something for fear that a certain someone will see it, then think twice about it.

If your friends have commented that your posts tend to push some limits or may anger or offend other people, ask yourself what you're getting out of sharing overly personal information. Is it really necessary for your boss to know about your weekend? Does your babysitter have to see you in your new bathing suit? Does your "frenemy" really need to know how you feel about her in a public forum? Ask yourself why you expose your every move publicly.

Take time to think through different possibilities in terms of how you express yourself and consider whether your life moments may be more appreciated and valued if not made so public.

We can't also share every detail of our life on Facebook and then complain when others share our status updates or photos, nor can we expect other people not to weigh-in on what we post. But, we also can't fully connect with others without disclosing at least some part of our lives. Wherever we choose to draw the line (and there should be a line drawn eventually) it is beneficial to keep in mind that not everything we do has to be shared. After all, there is at least one major advantage to not posting everything—once you stop worrying about how your thoughts, feelings and activities appear to other people, you can enjoy them more fully for yourself.

chapter four

How Many Friends Do You Have?

I probably know a handful of my Facebook friends. The others I met online. I don't know anything about them and they don't know anything about me. I prefer it this way—they know all about my life, but they don't know me.

Connecting in the Social-Media Age

"Friending" wasn't a verb before social networks like MySpace, Friendster, and Facebook existed. As if our daily social interactions weren't already complicated enough, Facebook has created a space where we're forced to redefine social terms that previously felt very stable.

Before social media, most of us met our friends through shared common interests or through introductions made by mutual friends. The depth of our connections rested on what information we shared and how often we shared it. While most of our casual acquaintances knew some things about us, access into the most private aspects of our lives was typically reserved for our best friends. But has the friending concept present in most forms of social media radically altered our views on friendship?

The simple act of friending someone grants them entry into your online life, for example, keeping up with your daily activities, knowing where you've "checked-in" at any given time, or knowing with whom you're talking. Whether you're ranting about your mother-in-law, posting pictures of your latest culinary creation, or sharing details about the adventures of online dating, friending gives a lot of access to your business. While it's common for your offline friends to receive a glimpse of what you're up to, on Facebook, many more people get far more than just a glimpse—and somehow we've grown to be okay with this. Is Facebook friending really about friendship? Although there's definitely an element of friendship found in online interactions, friending has more to do with fulfilling a validation need.

When many of us join Facebook, we discover people we know from every phase of life, sending and accepting friend requests enthusiastically. We may feel both celebrated and accepted. But then, we may find some anxiety setting in. *Am I friending too many people? What if some of them still keep in contact with people I want to avoid...for the rest of my life? What if someone I don't really like, a friend of a friend, friend-requested me? Is it a major faux pas to say no?*

How do you handle friending? Think about the three common groups of friends on Facebook: people whom we actually know, business or organizational friends, and complete strangers. Adding as many people to one's friend-list, without regard to whether we actually know them, is known as "hyper-friending" or "friend-whoring." Do you think you might be a hyper-friender? How many friends does a friend-whore make?

Inviting people we hardly know to be our "friends" is admittedly a bit odd. Typically, we don't invite complete strangers into our real-life world, so why are so many of us willing to share our most intimate thoughts and feelings with people we hardly know? Perhaps we feel safer interacting with people through a computer screen? Perhaps we're all looking for something more than friendship? Maybe many of us are lonelier than we think.

As humans we have a strong need to interact with each other. This need is what drives many of us to seek friends and followers

Max, 20

Detroit, Michigan

I love flirting on Facebook. I scope out my friends' lists every day for cute girls and try to add them to my own friend list. I only add hot chicks. If one changes her profile pic to an ugly photo, I hide her from my feed by deleting her comment, or if it's really bad, I unfriend her. I know I'm being shallow, but I have a system going. If women see a lot of other sexy women on my friend list, then they want to get with me. It's like a competition thing. I have over a thousand friends; most of them are hot women. I don't know them, but I need people to think I do. I want other girls to think they all want me.

◇ Like   💬 Comment   ➡ Share

on Facebook. Those of us who add as many friends as we can, do so because we want to feel important. We need followers—people who will look at our life and provide support or flattery. Initially, friending many people may make sense.

Friends on Facebook are different than our regular friends; they're our audience. If we have something to say, we no longer pick up the phone or text a friend with news, we proclaim it to our personal fan base. The problem isn't when we seek support from actual friends—that's normal. But when we turn to people who serve no other purpose but to provide us with unquestioning praise, our sense of self and friendships become distorted. This imbalance causes our "real" interactions to become affected. Inevitably, as your

Ted, 43

Boston, Massachusetts

A few months ago I called my friend Monica at work to ask her to please like one of my posts. She was seemed shocked and laughed at me and gave me some grief, but I stressed that it was very important and asked her to log on immediately. She asked why it mattered so much. I need at least ten likes on my posts. I only had five. Everyone loves my posts and for some reason I only had five likes this time.

Like   Comment   Share

list of friends grows to include your acquaintances, coworkers, the deli guy, and the "friend-of-the-friend-of-the-friend," the concept of friendship becomes somewhat cheapened and possibly lost.

During a face-to-face interaction, like a dinner with a friend, you can provide each other with compassion, support, inspiration and motivation as you drift away from the daily land of "shoulds" and "have-tos" to a realm of nonjudgment and limitless possibilities. Although developing an intimate friendship online can be done, it's difficult to conceptualize having this level of connection with someone on Facebook without it eventually materializing in real life. Even though Facebook has unquestionably extended casual contact with more people, most of our online casual friendships do not develop into close and personal friendships. Intimate friendships are complex and can require effort. In contrast, Facebook friendships, for the most part, are meant to be light in nature and are maintained in order to receive information or positive feedback and support.

On Facebook we reap the benefits of validation and praise without the social obligations of actually having to interact with all our Facebook friends in real life. Facebook gives us the opportunity to brag about having "1000 friends," but is that really something to brag about? We may be comfortable being friends with someone on Facebook, but sometimes we draw the line there and choose not to be friends with them offline. We've gotten comfortable with distance. By today's communication standards, even speaking to someone on the phone is becoming too time-consuming. Our real friendships are beginning to take a back seat to our online inter-actions. Even the way we describe our friends has changed, with phrases like "We're just Facebook friends," adding a new, almost casual dimension to friendship that didn't exist before.

Yarah, 32

Portland, Oregon

I became Facebook friends with Nina through a mutual friend. Nina seemed to enjoy my daily funny posts and posted similar things on my wall. Then she sent me a personal message on Facebook asking if I'd like to 'get a beer sometime.' I really enjoyed interacting with Nina, but I wasn't sure if I wanted to invest in a 'real' friendship with her. The thought exhausted me a bit. I already had enough friends and didn't know if I had time to spare for another, so I ignored the message and just continued my online banter with Nina, hoping that she would forget the whole thing.

🖒 Like   💬 Comment   ➡ Share

You may find that you start spending more time communicating online and less time making plans to actually meet friends out—in the real world. The way you communicate with friends may change too. Many people now spend less time talking on the phone and more time trying to elicit responses from each other through sharing witty commentary, funny videos, or quick "Happy Birthday" comments online.

I do believe that there is a space in our lives for online friendships that are healthy and better kept online, rather than bringing them into our offline lives. Many of them remain online friends because they offer us fun and casual humor, advice, validation and reassurance. These relationships are especially important when they present us with resources and conversations that we do not receive in real life, but what keeps them from evolving into offline friendships is the mutual level of self-disclosure shared.

As we develop a close friendship with someone we become increasingly direct and forward about our thoughts and feelings. We feel secure knowing that we can reveal who we are without judgment. As we unveil more and more aspects of our personal life, we assume that our friends will desire to do the same. This deeper level of self-disclosure is what connects people and makes it possible for one to, in a sense, adopt friends as a part of one's family. Mutual sharing is necessary in order to build trust. However, this can also backfire if we share too much information or if we share too much of ourselves too quickly. This "sex on the first date" behavior is rampant on Facebook. It's as if our personal profile has become a journal that just about anyone can access. Our sense of what is appropriate versus what is not is becoming blurred and our relationships are suffering because of this distortion.

When Facebook Friendships Affect Real Friendships

A recent study showed that one in five people has cut a friendship short after getting into a fight online. Many clients, friends, and other psychologists began sharing their stories of having to break up with friends because of Facebook. There is no doubt that many of us are interacting with our friends in a different way. Our Facebook interactions are taking precedence over real interactions, and some even try to solve problems through interpretations and assumptions made from Facebook posts.

Maria, 20

Houston, Texas

I had an argument with my friend Amanda. A few days later she posted a picture of friends on her Facebook wall, but she'd cropped me out of it. I was hurt and disappointed that she'd behave like this, but was more concerned that she tried to hurt me publicly. I felt I had no choice but to unfriend Amanda until we could work out our problems directly and in person, not on Facebook. Amanda couldn't accept that I needed a boundary. We were at an impasse in our friendship and couldn't work through it. The thought unnerved me. We had been close friends for four years, and our friendship was destroyed (in a matter of seconds) due to her posting a Facebook photograph that was intended to hurt me.

🖒 Like　　　　　💬 Comment　　　　　➡ Share

Friends are meant to be loyal. These are the people you turn to for comfort when you're going through a rough time. They speak constructively even when it hurts. They love you, not in spite of your faults, but because of them. Your friends have your back; they fight for you; they protect you. All the clichés by which we're surrounded exist for a reason. These real friends aren't going anywhere regardless of Facebook.

CONFLICT RESOLUTION

Before we take a look at how conflicts are dealt with on Facebook, let's first take a look at how we typically solve problems with friends in real life. Think of an argument you've had with someone recently. Chances are you felt a strong need to prove your point and show them how they're wrong and you're right. As much as we try to listen to a person's point of view, it's hard to do this because we're too busy thinking of our next line of argument. Our feelings can get in the way of thinking logically and sometimes we say things we do not mean. Hurtful words can add up and if we hold on to them, they fester and cause a lot of damage to the people we care about the most. Ultimately, in order to maintain a friendship, both parties should seek mutual understanding and compromise after they've both made efforts to work through concerns and repair the damage. Such discussion should happen in person or at the very least, on the phone, because working through a tough situation often requires us to see or hear another person's emotional reactions.

Now imagine trying to do all of this on Facebook. Your posts and written comments can be interpreted in many different ways. Because you can't hear a person's tone of voice, it's sometimes difficult to decipher the intention behind a post. Who has time for this anyway? Many of us know that Facebook is not an appropriate venue to take out frustrations on friends, but we're often the culprits, becoming passive-aggressive and not even aware of our behavior.

WAIT! HE UNFRIENDED ME?!?

Another dynamic specific to Facebook friendships is our ability to "build walls" around those who we choose to interact with and those we choose to ignore. In real life, when someone wants to stop seeing you, they just disappear from your life. Although painful, we get through it and move on with our lives. However, separations on Facebook often seem to be particularly painful. If you weren't expecting the unfriend, you are made instantly aware that your former friend no longer wants to know anything about you and they do not want you to know anything about them either. Unless the feeling is mutual, the resulting feeling is a deep sense of shock and rejection. Some people who are wounded from the unfriending will post hostile comments on mutual friends' walls so that the unfriender can see them. Some have even gone so far as to actually share online their version of the fallout as a form of payback.

Unfriending someone can have some nasty consequences. But what about those cases where there are no hard feelings and you've simply grown tired of their posts?

Many people are becoming bored with their Facebook friends or Facebook itself. It's very taxing trying to keep up with three hundred or more friends, and as much as people may whine about how much time Facebook is taking from their day, they may still feel compelled to check friends' updates.

Some decide that they need to limit their friend list. After all, how many friends does a person need? But they stop themselves out of a sense of obligation and a fear of hurting feelings. Most of the time simply unfollowing a friend or hiding them from your News Feed solves the problem of their posts intruding on your online world. Other times, a more extreme measure is required in order to maintain our sanity (more on toxic Facebook behavior in Chapter Seven).

THE RULES OF UNFRIENDING

On Facebook, are you guilty of excessive chatter? Do you post an occasional dramatic online rant or find someone else's entertaining?

Sean, 50

Jacksonville, Florida

During the presidential election, my friend and I were on opposite ends of the political spectrum and were bantering on Facebook. At first it was amusing but after a while it wasn't fun. I found myself feeling compelled to argue my own political points on my wall to refute his. I was even researching the subjects we were arguing about. I felt more and more anxious and angry over seeing posts I didn't agree with. Although I respected him, I couldn't tolerate seeing his posts and after some thought, decided to unfriend him. One day he called me and asked if I unfriended him due to his political preference. I admitted it, emphasizing we're better as friends offline. He agreed.

◇ Like   💬 Comment   ➡ Share

Reasons people choose to unfriend others:

🦃 Inappropriate posts that share too much personal information: Conversations better suited for private settings are discussed openly, forcing friends to share in bedroom romps.

🦃 Political or religious affiliation: No matter what you believe or stand for, you're not likely to convince anyone to join your team.

🦃 Facebook drama: Why go to the movies when there's this much scandal on Facebook for free?

🦃 Excessive negativity: When people feel horrible about themselves, they want others to feel horrible too.

🦃 Excessive optimism: She worked out in the morning, fed homeless kittens in the afternoon, and then did charity work for four hours after work. No one likes a show-off.

🦃 Posting too many profile pictures: Chances are, she's either newly single or showing her friends how hot she truly is.

🦃 Excessive chatter: People don't have the patience to read a three paragraph explanation of all of your insights on life. Also known as TLDR (too long, didn't read).

🦃 Sharing too many quotes: Boring.

🦃 Meaningless updates: No, we really don't care what you purchased at the grocery store.

🦃 Meanness: Friends are supposed to be supportive and loyal.

Or off-putting? Most of us have had "venting" days. Are you feeling compelled to share too much? There's a lot of one-upmanship on Facebook, isn't there? All too easily we get caught up in how we appear online, forgetting that we don't need to share everything that crosses our mind.

TO DELETE OR NOT TO DELETE?

This is the question one is faced with in real life as well. Unfortunately, Facebook can make this question reappear in our minds almost weekly. Aside from negative online interactions, people also unfriend due to a friend's behavior in the real world. Perhaps you overheard a friend insult your partner's looks or a coworker complain about your spreadsheets to your boss. Other factors affecting whether to unfriend are geographical differences and how much people value their friendships before unfriending. If you weren't good friends before, chances are when someone upsets you it will be really easy to walk away from them. Regardless of why we unfriend people, we often don't approach letting go of friends the same way we would if we were face to face. In theory, in real life we should speak openly and directly to our friends about our concerns. On Facebook we will hide, unfriend or block someone without discussion. When we unfriend someone the consequences can carry on to real life.

WHERE DO WE GO FROM HERE?

Facebook friends provide us with laughs, entertainment, a nice distraction from our daily pressures and, most important, connectedness and a feeling that we belong. Facebook friends can give us much-needed support and validation when we may lack these things in real life. However, our real-life friendships offer us moments and experiences that simply cannot be replicated online, like watching your friend's face as she laughs at one of your jokes or the profound compassion you feel when your friend discusses a stressful experience with you.

Although our online interactions can be wonderful, they should not be our main interactions. Studies show that because of increased social-media communication, we are spending less and less time with our friends in person. Unless there is a specific reason why we are limited in our real-life interactions, we should make efforts to spend more time with our friends in social situations.

Additionally, Facebook conversations are too easily misunderstood and misinterpreted. Such misunderstandings will inevitably affect our real relationships. Try to find a balance between how much you interact with your real friends and online friends. When a friendship has become too painful to be maintained, sometimes it's wise to remain silent. When you speak in frustration or anger, your point is often lost. Knowing when to refrain from speech takes wisdom and a deeper understanding of what is conveyed through silence.

Additionally, silence sometimes communicates more than actual words. Think of that stern look your mother used to give that communicated "stop it" more than words ever could, or the message you clearly received when someone ignored you. Silence is an indicator that you are above a petty argument. Take the time necessary to cool down and then make efforts to speak to your friend in person about your concerns. Set some boundaries for yourself if you ever find yourself in this situation in the future. Make an agreement that you will not respond until the following morning. You may find that doing so can mean the difference between resolving a misunderstanding and ending a friendship that you truly valued. Think of the friend or two that you've lost because of a silly misunderstanding, a moment of reacting too soon or doing something online that leaves you wishing you had a do-over.

This common situation holds true on Facebook: Treat others the way that you wish to be treated. Try the new Golden Rule: If you wouldn't say it in real life, then don't say it on Facebook.

chapter five

Relationship? 'It's Complicated'

I lost myself through spending too much time with his Facebook crap. We were dating for five years. He kept playing games with me and I fell for it. He'd flirt with other women on Facebook and then call me paranoid when I brought it up. Flirting on Facebook is no different than flirting in real life.

LOVE, ROMANCE, AND FLIRTATION ON FACEBOOK

WE SHARE A LOT OF THINGS ON FACEBOOK: WHAT WE'RE thinking, where we're going, what we're doing, and every once in a while during a spark of romantic sentimentality or frustration, we may even share how we're doing in our relationships.

Expressing feelings about our relationships on Facebook can be a nice way to connect with our partners. It seems that most people revel in their intimate relationships online. I'd even go so far as to say many love it and take personal pride in it. Whether romantic photographs or notes of endearment, a romantic posting on Facebook is often one of the strongest expressions of status happiness. These postings can be enjoyable not only for the couple, but also their friends, becoming connections of optimism in the simplest form.

Then again, some people find romantic exchanges on Facebook unpleasant and, in some cases, unbearably so. In fact, relationship turmoil triggered by Facebook is a very common topic in psycho-therapy—interacting with a former flame, obsessing over your girlfriend's flirtatious picture, over-analyzing interactions on your boyfriend's wall—all can be major issues in romantic relationships. Add the growing trend of people altering and embellishing their

Facebook profiles in order to trick someone into falling in love with them, and you have the makings for true soap-opera level messes.

In real life, when you meet someone for the first time they can easily lie about their achievements, their girlfriend, their children, or any other truths about themselves. Facebook, however, has created a platform where lying is even easier; it allows any of us to present the best parts of us without exposing the bad. It's no wonder people often get caught up in online relationships that are based on illusions. Think about it—let's say you meet a total stranger on Facebook. How would you know that their profile is a true representation of who they really are? How would you know that this person is actually who they claim to be?

CATFISHING: THE NEXT GENERATION OF FRAUD

Catfish: The TV Show is an MTV reality-show series about the truths and lies of online dating based on a documentary movie of the same name, which focuses on how people present themselves on their Facebook profiles. Catfish is a term the series uses to describe people who create fake profiles on social-media sites pretending to be someone more appealing than they are in real life. They share false photographs and biographical information and, in general, forge new identities from which to live.

In the show, the host and executive producer, Yaniv "Nev" Schulman, along with his filmmaking partner Max Joseph, help couples meet for the first time. All of the couples featured on the show have interacted only through social-media sites, sometimes for months or even years. The show travels to where one person lives, conducts an online investigation of the online profile in question through Google or other search engines, and then makes contact with the other person in order to arrange a meeting. The entire first-time meeting is filmed.

The main allure of *Catfish* is intended to be the anticipation of figuring out whether or not the love interest has created a fake online profile, whether or not this individual is "real." A secondary hook to the show involves the hosts examining why someone would create a fake profile. In most cases, people create fake profiles because they're trying to hide some insecurity they have about themselves. In some cases, the profile turns out to be legitimate. In others, the online identity is discovered to be a sham. This discovery leads to feelings of betrayal and heartbreak for the person who fell for the fake profile and the whole encounter ends up in disaster.

This show also reveals the feelings, including the emotional rollercoaster, of those who "catfish" and those who are the object of the catfishing. One *Catfish* episode in particular intrigued me: The story of Ashley and Mike (Season 2, air date: 9/03/13).

For more than seven years, Ashley sent dramatically altered photos of herself to Mike in order to make her look thinner and more attractive to him. Mike's profile photos show him as an extraordinarily fit and attractive man. Ashley tells the hosts that she altered her photographs because of her insecurities about her weight and a fear of going out in public. Although she felt emotionally connected to Mike, Ashley kept finding excuses not to meet up with him. By the time Ashley contacts the hosts, she feels that she is ready to finally meet Mike (in person) and reveal her true self. Ashley describes her situation to the hosts:

I met Mike when I was 13 in a chat room online and I started, since then, Photoshopping my photos because that's how I thought I should look in front of the world. When I feel like

it's gross, or a double chin or, you know, a muffin top or thick
legs, thick arms, I just adjust that to make it more comfortable
for me to post for everybody in the world to see it...now it's
been seven years and it's kind of hard to tell him casually
over the phone, 'Hey, I Photoshopped all my photos.'

Ashley describes herself as a hermit—she avoids spending time
in public in order to avoid the negative comments people make
about her weight. Because of her negative experiences in the past,
she believes that she should be different than who she really is.
This belief, based on a deep-seated insecurity, is the motivation
to alter her online image. She believes that if she reveals who she
really is, she wouldn't be worthy of love.

On a certain level, we all wish we were a bit different than who
we are. Insecurity initially develops when we haven't received the
emotional validation necessary to develop a sense of self-acceptance and self-confidence. When we don't feel "good enough," we
try to hide our insecurities from others. Taken a bit further, many
also feel the need to exaggerate accomplishments or embellish who
they really are in order to feel more accepted.

When they finally meet, Mike is very happy to see Ashley and
tells the show's hosts he thinks she looks beautiful. He is happy and
enthusiastic to finally be able to hug Ashley in person. However,
Ashley's demeanor immediately changes from happy to cold and
distant. She avoids eye contact with Mike and responds to him
using mostly one-word answers. Mike notices Ashley's discomfort
and tries to comfort her, but he is met with rejection from Ashley.
She asks to leave his home.

The hosts speak to Mike about how he had presented himself to
Ashley online. He admits that he shared the photos from a different
person on his profile but figured that 'Ashley's a smart girl and

thought she figured it out.' He adds that he did not care that Ashley edited her own photos. He still thought she was beautiful.

Shortly following their first, in-person encounter, Ashley tells the hosts that upon seeing the true Mike, she no longer feels any chemistry between them. She blames her change of heart on Mike's deception, but admits she's being a hypocrite. She tells Mike that she is not ready for a relationship.

Any of us can appreciate the whirlwind of emotions involved with meeting someone online, falling in love, and then meeting them in person for the first time. Most of us can also relate to how it feels to be confused and disappointed by love. Sometimes the road to love involves heartbreak and deep reflection, but if we can face our vulnerabilities and allow someone into our world while remaining true to who we are, we might just end up at someplace unexpected and wonderful.

Ashley and Mike had a lot of soul-searching to do—something that's clearly not possible over the course of a single episode. They both had to take time to explore and understand the source of their insecurity. This episode ended on a promising note: Ashley took some time to face her fears, heal the wounds from the past, and learn to love herself. It was only after she accepted herself for who she really is, that she was able to reflect on everything that had occurred between her and Mike. She came to realize that she truly cares for Mike and wants him in her life. Mike in turn faced his own fears of being judged and learned to accept and appreciate his true self. They announced that they rekindled their feelings for each other and were making plans to reunite in person. The two did not end up starting a relationship, but remained friends.

The story of Ashley and Mike is an example of how two people can fall in love with an altered profile on Facebook. It also shows why people alter their profiles and the effect this has on both parties.

The lingering feeling of shame or embarrassment associated with falling in love with a fake profile resonates with many people, and is one of the reasons why this show has become so popular. *Catfish* shows us that not only almost anyone can be fooled by an online profile, but also that it's possible to fall in love with a profile—a created illusion that, for the sake of the heart, many are willing to believe.

But She's Just a 'Friend'!

Both men and women can be guilty of checking out other people's profiles and comparing them to their current partner, regardless of how happy they may feel in their relationships. Facebook creates a world where pics of people can make us susceptible to unrealizable wants and hopes. Many attractive people are "in your face," so to speak. This temptation may lead us to post flirtatious comments on their wall, which can cause our partner jealousy, triggering an impulse to dig deeper into what's going on. We know that adding an attractive stranger to our friend list may make our partner jealous, but many people can't help themselves; the curiosity of finding out more about this person is the hook.

For some, finding an attractive Facebook profile picture can be intriguing. They may wonder how that person will react to them (or to their Facebook persona, to be precise), and if flirtation follows, they feel good about themselves. This kind of Facebook interaction can be an immediate ego-boost for individuals who really need one. Friending attractive people can be strategic too. People who do so may feel that having so many beautiful friends makes them the Casanova of Facebook, as if their attractive friends are a reflection of their own attractiveness by association.

Kerry, 32

San Francisco, California

I met this guy on Facebook and I dated him for seven months. He was gorgeous and on Facebook he made himself out to be so amazing. He posted love songs and wrote poems for me on his wall. We talked for weeks and then we met in person. I fell so hard for him. But I didn't know what he was capable of. On my birthday I didn't hear from him at all, which was weird. Before then, we spent every day together. While I was at my birthday party, one of my friends called me to tell me that I better check my Facebook wall— NOW. This asshole posted a pic on my wall of him making out with another girl, and captioned it: "Happy Birthday Sweetheart!" I just stared at it in disbelief. I couldn't move and I couldn't breathe. All I thought was that this must be some kind of mistake or joke, but it wasn't. He really did that to me. He broke up with me on my wall by posting a pic of him making out with someone else. All my friends started calling me and shouting that they couldn't believe it either. I wanted to tell him off, but I didn't. Sometimes these jerks just want to hurt you and they want to see you cry. He wasn't getting that from me, so I did nothing. I completely ignored it and left that picture on my wall. I figured if he wants to make himself look like a jerk, why stop him? I found out later on that he married that girl and now she cheats on him left and right. Now he wants me back. Ha! Good luck with that you loser.

👍 Like   💬 Comment   ➡ Share

Such interactions are no doubt seductive, but are mostly based on illusion. When we act on them, we flirt with disaster. Many people are aware that through these behaviors they are risking their own relationship, but the intoxication of Facebook interactions can be too strong to avoid. How much can Facebook affect a relationship? Divorce attorneys commonly use Facebook flirtations in their cases and even joke that they should solicit new clients on Facebook itself. NBC Bay Area News reported that Facebook is cited in 20 percent of divorce cases, and 80 percent are using social media to connect with lovers. I've witnessed friends worry over how many attractive men or women their partners have friended or

Kristen, 33

Concord, New Hampshire

Matt and I have been dating on and off for over five years. We broke up in college but we recently reconciled and our relationship is stronger than ever—except for one problem: Matt hasn't changed his Facebook relationship status to 'In a Relationship' or mentioned me at all on his wall. Matt is vague about why he doesn't want to change his relationship status, which is making me suspicious. My friends' boyfriends mention them on Facebook. So why can't Matt? Aside from this, our relationship is great, but I strongly believe that if you're in a relationship, you should be willing to announce it to all your friends. I think Matt is hiding me—from someone else.

☺ Like   💬 Comment   ➞ Share

torture themselves wondering why their spouses are flirting with old partners. There's even a Facebook page called, "I wonder how many relationships Facebook ruins every year," with over 20,000 "likes." In my own practice, I hear about this topic almost every day.

The perception of how others view our relationship should never become more important than our own perception of our relationship. If Kristen and Matt are happy in their relationship, why does she care about a social-media status update? Because, for some, announcing a Facebook relationship carries more weight than the happiness of the relationship itself. Facebook shows us how everyone feels about their relationships, and sometimes, we can't help but compare ourselves to others.

Psychologically, we get our self-esteem mostly from the perspectives and influences of others. When people change their relationship status or post pictures of their beloved, we may feel snubbed if our partner doesn't do the same. For some, the acknowledgement of a relationship represents how the couple values each other, even more so than their real-life displays of affection. Somehow the real status of a relationship doesn't seem "legit" until it's proclaimed publicly through Facebook's relationship status feature.

The relationship status is arguably one of the single-most powerful Facebook features and tends to cause a lot of emotional uproar. People spend days, weeks, months obsessing over someone's relationship status. I've even heard a few people refer to this Facebook feature as the "curse"—the minute someone changes their status to "In a Relationship," the relationship in question is surely doomed. Some of our friends are changing their relationship statuses as often as once a week. What does this say about their love life? Once it's out in the open on Facebook, some people begin to wonder if they really want to be in the relationship after all.

Mike, 41

Fresno, California

I had been dating my partner for three months when we talked about being exclusive. Without my knowing, my boyfriend changed his relationship status to 'In a relationship with . . . ' and tagged me. I didn't know exactly how to feel about it. I was happy in my relationship yet felt somewhat uneasy. I would've preferred that we talk about it, that's all. Now I feel pressured to add him and I'm just not that comfortable making my relationships public. Plus, I would have liked to have known that we were totally exclusive before my friends found out! I eventually changed my relationship status too. But I asked him to discuss things like this privately with me before we mutually decide to make them public.

     👍 Like   💬 Comment   ➡ Share

LOVE ISN'T COMPLICATED, PEOPLE ARE

An interesting choice for a Facebook relationship status is "It's Complicated." This option allows you to instantly tell your friends that you're happily—or not so happily—involved, or that you're single once again. This option is also an instant and dramatic way to get attention from your current partner. On a more subtle level, it's code for "I'm in a relationship, but am willing to cheat." Why would someone publicly announce something so private? Because

Facebook makes drama easy. If the relationship status feature didn't exist, we wouldn't be able to broadcast our relationship problems so easily. Before Facebook, our relationship drama was kept between us and our closest friends. Now claiming that our relationship is "complicated" is a teaser, inviting comments by concerned friends and gossip from others.

Our relationship status also may have a direct influence on the direction of the relationship itself. If a couple breaks up on Facebook, then gets back together, things don't just move forward from there. Everyone knows there was a problem and wonders what else is going on. It's as if you're allowing others around you to become a backseat driver to the relationship; the couple may feel the need to explain publicly what, before Facebook, would have been handled privately. Questions about the breakup may continue to haunt the couple and may lead to further problems down the road.

KEEP YOUR PERSONAL LIFE PERSONAL

People on Facebook are not relationship counselors, and your wall is not meant to display a play-by-play of the problems the two of you are going through. It may be tempting to vent about the frustrations you feel in your relationship, but rather than blasting them onto your wall, pick up the phone and call your best friend. Go for a walk. Exercise. Meditate. Do anything but log onto Facebook. Facebook is a public forum and not a personal diary.

Voicing your problems about your relationship to your Facebook friends may provide an immediate tension release, but will likely cause more damage in the long run. Working things out with your partner without Facebook's help demonstrates mutual respect and a desire to be strong from the inside out.

Many people have experienced extreme romantic behaviors played out on Facebook. The big contenders are: jealousy, stalking, obsession and revenge. In terms of jealousy, you may find your ex posting photographs of themselves with their new partner, or draw faulty conclusions based on who he or she has friended.

More often than not, we experience pain from our own assumptions. When someone posts something it can raise hundreds of questions: Was that about me? Is she trying to tell me something? Was that a poke at me or someone else? Who's that guy she's with and why is she posting pics of him all over the place? Is he trying to get me jealous? Questions triggered by Facebook posts lead to assumptions, assumptions can lead to painful emotions and our emotions, in turn, lead to actions that we would not otherwise take—like stalking.

FACEBOOK: HELPING STALKERS SINCE 2004

Stalking has never been easier. Prior to Facebook, in order to be considered a "stalker" you'd have to participate in highly inappropriate behavior such as waiting outside someone's house, calling mutual friends to try to gain more information, or showing up "by accident" at an event just to run into an ex-boyfriend or girlfriend. Now all you have to do to stalk an ex is turn on your computer, make a few clicks and navigate their wall, or a mutual friend's wall. Paying attention to where your ex is "checking-in" can reveal a pattern, allowing you to bump into them at bars and social events. Joining the same groups they follow provides opportunities to interact with them online even when they've asked you to leave them alone. Stalking easily leads to obsessive behavior where every single move and update is tracked and interpreted:

every "friending," new photograph or tag gets analyzed. In extreme cases it can lead to true stalking and create a situation where the person being stalked feels unsafe (more on this later).

Similarly, suspicion and jealousy can make people behave irrationally and plot revenge. Posting something hurtful on Facebook harms some people while they're trying to cope with their loss.

Edward, 44

El Paso, Texas

My girlfriend, Samantha, broke up with me suddenly, saying she needed a break. She told me I was too controlling and that I needed to work on myself for a while. I hated the idea of a break and felt really anxious when she did not respond to my texts. She said that she can put up with a lot in a relationship, but when she's done, she's done. I freaked out. Did she just want time apart or did she say that as a way to get rid of me? I got pissed when I noticed that she posted a photo of herself with another guy. I needed to get even! I called an ex and took her to dinner, planning to post photos of us hugging on my wall. I knew Samantha felt insecure about this ex-girlfriend and knew it would hurt her. When she saw my photos Samantha completely broke it off with me. I found out later that the man in Samantha's photo (the one that triggered all of this) was her cousin.

☃ Like   💬 Comment   ➞ Share

Samantha clearly needed a whole lot of "gone" between her and Edward. Whether it's online or offline, if you find yourself continuously trying to explain to your partner how they are disrespecting you and they keep doing so, it's time to move on—period. There's no way that you will ever be able to attract mature love, respect, and trust into your life by allowing yourself to become someone's emotional punching bag. Do whatever you have to do to remove toxic people from your life—unfriend, block, get a restraining order—whatever it takes, forget about them, and move on. Post break-up, don't engage them online, at least not until you feel that you are in a better place emotionally. Believe me, toxic individuals will share almost any provoking post or photograph to elicit a reaction from you. Don't fall for it. If your ex respects you at all, they will honor the fact that you need time alone and space away from them.

In the real world it is a lot easier to avoid such painful situations, like avoiding the restaurants and clubs you and your ex used to frequent together. On Facebook, however, avoiding your ex is almost impossible without unfriending or blocking. Constant exposure to what your ex is doing (or who they are doing it with) becomes a way to retraumatize yourself, or at the very least prevent yourself from moving forward. Sometimes your ex may purposely post a photo or tag you in a photo of them and their new love in the hopes of making you jealous, while other times Facebook-induced jealousy can take you completely by surprise. *The Huffington Post* shared an article, under their Weird News section, about a man who assaulted his wife because he thought she was flirting with a man on Facebook:

Tennessee man Lowell Turpin became jealous after seeing his live-in girlfriend of five years, Crystal Gray, in a photo with a man he didn't recognize. She seemed to be following this man on Facebook regularly. After accusing her of having an affair, he smashed her laptop against a wall and proceeded to physically assault her. He was arrested on domestic battery charges. He hadn't recognized the man as Republican presidential contender Mitt Romney.

This is an extreme, and slightly hysterical, example of Facebook-related jealousy, but data shows that more than 70% of Facebook users follow the activities of their exes regularly, sometimes going so far as to "friend" their ex's new partner. Some even go so far as creating fake profiles to maneuver their way back into their ex's Facebook life after being blocked. Ultimately, if you're still looking on your ex's wall, you're not making progress toward a brighter future without them.

Nobody likes to think that their ex has "traded up" on the dating scene, and though Facebook lurking can be a source of great pain, some people just can't find the will to stop. In addition to jealousy, we can experience profound sadness, anger, fear and anxiety as we follow someone's posts.

Unfortunately, Facebook has also added to the long list of ways people can break up a relationship. Breaking up through a text message is bad enough; breaking up on Facebook adds the dimension of being public, like being broken up with in the middle of a crowded shopping mall filled with people you know.

People who remain friends with exes on Facebook may have a lower capacity to get over it and move on than those who do not remain Facebook friends. Constantly lurking around an ex's Facebook page may lead to higher levels of distress about the

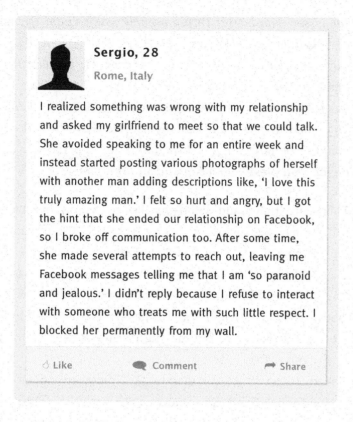

Sergio, 28

Rome, Italy

I realized something was wrong with my relationship and asked my girlfriend to meet so that we could talk. She avoided speaking to me for an entire week and instead started posting various photographs of herself with another man adding descriptions like, 'I love this truly amazing man.' I felt so hurt and angry, but I got the hint that she ended our relationship on Facebook, so I broke off communication too. After some time, she made several attempts to reach out, leaving me Facebook messages telling me that I am 'so paranoid and jealous.' I didn't reply because I refuse to interact with someone who treats me with such little respect. I blocked her permanently from my wall.

⌀ Like   💬 Comment   ➥ Share

breakup, lowered self-esteem, continued sexual longing for the ex, and difficulties in growing beyond the breakup. It's important to note that I don't believe that Facebook causes these issues; it's more likely that people who have a hard time moving past a breakup will use Facebook to keep tabs on their ex and wallow in their sadness over the breakup. For these people, the best remedy for their heartbreak may be a complete disconnect with their ex, both on Facebook and in the offline world. And sometimes a helpful way to move beyond a failed relationship may be to have a temporary breakup with Facebook as well.

Deanna, 35

Las Vegas, Nevada

My ex-husband 'wall-watched' me throughout our divorce so much so that I eventually felt forced to deactivate my account. No matter what I did, he kept getting information about me. I couldn't figure out who was leaking information back to him. Mutual friends told me that he was acting strangely on Facebook, posting old photos of us during a happier time, changing his status back to 'married,' and announcing on Facebook that we had reconciled. A bunch of romantic quotations shortly followed. I started getting calls and texts from friends congratulating me or asking me what was going on. It's actually pretty funny now, but it really made me angry then.

👍 Like   💬 Comment   ➡ Share

WHERE DO WE GO FROM HERE?

There are two things you actually have control over on Facebook: what you post, and how you react to other posts. What you post about your relationship should rely on good judgment, communication with your partner, and a clear understanding of what's appropriate and what's not. If you have all three of these components, you can minimize the drama and disconnect. Avoid posting for attention or to provoke, and be genuine.

Every relationship is different, and while you can't control how people see your relationship you don't have to react to their comments. Admit it: Our definition of a "friend" on Facebook can be very different from one in the real world. Many of your Facebook friends, though they may say otherwise, may not have your best interests in mind when commenting about your love life—or anything else, for that matter. Keep this in mind when on Facebook and think about what your friends' intentions may be when reading their posts. Know who on Facebook "has your back," and who has other motives in mind.

Facebook was designed as a social network to bring people together, not tear them apart. Relationships can't survive without cultivating and developing roots for trust. If Facebook is causing issues, talk it over. Have a conversation with your partner about your friends. Tell them who—of your hundreds of Facebook friends—is important and who's not so important, and why. Put them at ease if any feelings of jealousy crop up. Keep direct and open communication about what you both agree to share or not share. The same goes for changing your status or friending certain people. It's best to talk about it first in order to avoid misunderstandings. These misunderstandings can cause our worlds to fall apart.

chapter six

Teen Cliques and Clicks

No conversation is private anymore. If you text something, your friends will take a screenshot of your conversation on their phone and forward it to everyone else and then it'll get posted on Facebook or Instagram. And once it's out there, they'll never stop harassing you. They won't let it go. Your life gets ruined.

Facebook's Unique Challenges for Teenagers

ONE SUNDAY AFTERNOON, MY FRIEND'S TEENAGE DAUGHTER and her classmates sat down with me to discuss their social-media interactions. Most of them connect through social-media sites including Facebook, Twitter, Instagram, and Tumblr. Social media has been taken up by American society of practically every age, but teens take online connection to a whole new level. At the drop of a Facebook friend, these kids can tell me who they're talking to, on which site, and at what time. On a daily basis, some teens check in with their 1000-plus Facebook friends, 300 Twitter followers and 70 Tumblr bloggers. They can't imagine a life without social media.

Over the course of six months, I sat with my friend's teens, my teen clients, and other random teenagers who agreed to an interview. What I discovered is that social media is as integral to the modern-day teenager's life as home telephones were to teenagers in the 80s. According to the May 2013 Pew Research Center's report on U.S. teens, 81 percent of Internet-using teens reported that they are active on social-networking sites, more than ever before. Facebook, Instagram, Twitter, and new dating apps like OkCupid, Tinder, Blendr, and Grindr have become key players in their social-media interactions.

Teens have unique challenges and behaviors on social media, and in this chapter we'll see how teens interact with each other through social media and how it affects their identity development, social skills, attention span, flirting behavior, and their ability to cope with cyberbullying.

'Digital Natives'

Active teen social-media users meet people through social media, flirt on dating sites, and spend hundreds of hours online. We couldn't have imagined this world even twenty years ago. The new generation of teens (also known as "Digital Natives") are not only social-media masters, but are also ultra-tech savvy in many facets. Just how many tech-savvy teens exist today? The Pew Report shows that 37 percent of American teens ages 12-17 now have a smartphone (up from 23 percent in 2011), 23 percent of teens own a tablet, and, overall, teens represent the largest group connected to the Internet.

Today's teens prefer texting to phone calls as a way to keep in touch. They have grown up with iPads instead of books. They text on skateboards, weaving in and out of traffic, leaving their parents perplexed and worried. Is all this new technology somehow stunting teen development?

Teens using social media to interact may not be so different than some adults, but there is one glaring difference between the generations: today's teens have grown up with smartphones and social media, and a large majority of today's teens understand themselves and each other through these devices. What does this wave of super-connectivity mean for their psychological development?

Psychologist Erik Erikson's *Theory of Psychosocial Development* is one of the best-known theories of personality. Erikson believed that our social experiences shape our personalities in a series of stages. One of the main elements in Erikson's theory is the development of our sense of self, known as our *ego identity*.

Psychologically, during our adolescent years the main theme of life is *Identity versus Role Confusion*. During our teen years, we develop and shape our sense of self. Teens successfully form their identities when they receive proper encouragement to explore their environment in an independent way. Those who aren't able to form a sense of self remain unsure of their beliefs and will likely feel insecure and confused about themselves and the future.

Our sense of self constantly changes due to new experiences and information we acquire on a daily basis. As with their offline interactions, computers, smartphones, and social media have substantial influence on how teens view themselves, each other, and the world.

Smartphones give us instant access to the Internet and each other. For those of us in our twenties, thirties, or forties, this means we possess a capacity for communication unseen in the past. For teens, the benefits are the same; however, such instant accessibility can also mean a possible radical shift in Erikson's development theory. Adults using technology today have already developed their ego identity, while teens are still in the process of shaping their sense of self.

Smartphones are flooding teens with information, which not only over-stimulates their senses, but also affects their ability to function in an organized manner. The more teens interact with each other online, the more likely their ability to absorb and process information will become affected. This can lead to a low tolerance for delayed gratification, increased impulsivity, poor coping skills, and even symptoms of Attention Deficit Hyperactivity Disorder (ADHD).

Many of the teens I interviewed displayed poor social skills. Throughout an hour-long interview, they interrupted each other, got lost in conversation by jumping back and forth between topics, and appeared to have difficulty maintaining focus. They had a hard time maintaining eye contact with me, and each other, preferring to focus their gaze on their smartphones.

Some teens treat their mobile devices like an extension of their body—always within arm's reach, and when they're not texting they're scrolling through social-media sites like Facebook, Twitter, Instagram, Tumblr, and deviantArt.

It's Not All About Facebook

According to Bianca Bosker, Tech Editor of *The Huffington Post*, most teens are still on Facebook but they view it the same way as adults view email—they don't love it but use it to communicate. During a *Huffington Post* video interview, Bianca tells us about teen Facebook use:

> *Facebook has become the living room of the Internet. Your parents are there, your teachers are there, your classmates, the hundreds of people you might have met at camp. Twitter and Tumblr are increasingly sort of the basement rec-room where you can have your own conversations; it's more intimate, which is very interesting 'cause it's not like Twitter is a small thing. There's hundreds of millions of users, but teens feel like they can have a little more privacy and avoid all the drama.*

Today's teens are not confining themselves to Facebook but are using other social-media sites where, through multiple or faceless accounts, they can express their ideas and images with more anonymity. The Pew Study says that most teens who use social media (94 percent) have a profile on Facebook, but they are simultaneously integrating to other sites such as Twitter and Instagram.

Many of the teens mentioned "the drama" as a key factor in opting out of Facebook. They prefer other networks where drama is not so prevalent and which are parent-free zones.

Despite complaining about drama or losing enthusiasm for Facebook, most teens aren't deactivating their accounts. What keeps them logged on is perhaps a phenomenon known as FOMO (Fear of Missing Out) on important information, social events and interactions. Since most teens still interact through Facebook, many teens

Melissa, 15

Chicago, Illinois

I just got so sick of Facebook. It started out okay. I used to love Facebook, but then people got crazy. People would share pics and compete over who looks better or who's doing cooler things. My friends would talk shit about each other on their walls, but the worst of the worst was when my mom started acting crazy on it. She took tons of 'accidental' selfies in bathrooms. Then she'd check in at the places my friends were hanging out at. She kept posting and tagging me on stupid pics. It was so embarrassing. I'm still on it [Facebook] but I like Twitter more, where my friends act normal with zero drama.

    ♡ Like   💬 Comment   ➡ Share

feel obligated to stay logged on. However, other social-media sites are quickly gaining the attention of teens as a viable means to get their messages across.

According to a 2012 McAfee study examining digital activity across multiple computing devices, 70 percent of teens actively seek to hide their online behavior from their parents. Most teens I interviewed revealed that they are on Facebook, but not really on it. Meaning that they have a Facebook account that they don't use, but keep it as a sort of decoy. So while parents monitor their teens' Facebook accounts (because many adults are also on Facebook), teens turn to other social-media platforms to express themselves more freely. For this reason, most teens do not restrict their parents from their Facebook profiles. Similar to how rave culture in the 90s used post systems at public places such as record shops, today's teens are also experts at finding places to secretly meet and express themselves.

Regardless of how teens view such sites compared to Facebook, many of these sites encompass similar expressions of drama. For example, thousands of teenagers can easily have "Twitter wars," which can be even more dramatic than Facebook because they are not relegated to just friends but the entire school using hashtags.

Some other social-media sites not mentioned in the Pew study are Pinterest, Vine, Reddit, Snapchat, Kik, and 4Chan. New sites built by and catering to today's teens appear every day—it's almost impossible to keep track of them all—and while they all have different features, a majority appear to emphasize photographs, video and audio as forms of self-expression. What do these new platforms afford teens in terms of self-expression and self-identity?

Summary of the Top 5 most popular social-media sites (in order) used most by teens according to the Pew study:

Facebook: Teens feel a need to maintain a profile, regardless of how often they're actually on it. The Pew study reveals that Facebook contains the most active teen users. The typical teen user has 300 Facebook friends.

Twitter: This site offers quick connection with anyone in the world, including celebrities. Users post updates in 140 characters or less.

Instagram: This social network for uploading and sharing photos is super popular with teens who like to upload images and selfies.

YouTube: Teens who use YouTube can update videos of themselves and their friends to their accounts and allow others to view their content.

Tumblr: A social network for short posts enables teen bloggers to share photo, video and audio posts that are often re-shared from other sites with minimal text.

THE NEW SELF-PORTRAIT

MySpace likely launched a widespread phenomenon known as "selfies" (aka: "bathroom pics")—photographs we take of ourselves with the intention of posting them on social-media sites for others to see. This tendency then spread to other social-media sites such as Facebook and Instagram. Some teens post selfies on social-media sites as a form of self-expression or as a way to stay connected to friends. Sharing selfies can be empowering for teens—showing them being involved with life and encouraging them to be a part of their environment. In the same breath, some teens post selfies out of insecurity. They post them in order to see how many likes (or endorsements) they can get from their friends.

Sandrine, 15

Schaumburg, Illinois

I've seen people post selfies when they're drinking or smoking weed. That gets them more likes on Facebook. Especially girls. They'll post pics of themselves looking slutty so they get likes. My cousin posted this one pic of him puking on himself at this party. He was on the lawn, like passed out. A dog could've walked by and pissed on him too. But it got him a lot of likes. Girls get naked; boys get high. That's what gets you attention.

🖒 Like   🗩 Comment   ➡ Share

According to research conducted by Dr. Larry D. Rosen, professor of psychology at California State University, Dominguez Hills, teen Facebook users are more likely to display narcissistic tendencies. With a platform designed to allow users to put a spotlight on themselves, social media encourages a focus on self-image like never before. One of the key benefits of social media is that it helps teens discover themselves through self-expression. Another viewpoint is that it encourages teens to focus on themselves too much. Every day teens are sharing their thoughts on Tumblr, their family secrets on Twitter, or sexuality confusion on deviantArt. Whichever form a teen's self-expression takes, social media makes sharing it instantaneous and widespread. Such freedom can lead teens to gain a deeper sense of self-worth—empowering them to act independently through their own choices. At the opposite end of the spectrum, they may feel invincible and this notion can give them a false sense of entitlement.

According to The Pew Report, teens strive for Facebook likes regardless of where the likes come from and seek followers on Twitter. On Twitter, people tend to follow other Twitter users with similar interests, so the more followers (or "fans") a teen has, the more likely it is they will believe that they're exceptionally interesting. Whether a teen receives likes or followers, such online endorsements may lead them to feel a heightened sense of self-importance. Sooner or later they will have to log off—even for brief moments—and face their offline reality. When they realize that they are not as interesting, important, or powerful offline as they are online, their world as they know it will come crashing down. Not being able to manage their offline identity may lead to anxiety and identity confusion.

Teens have time to edit and enhance their profiles in the privacy of their own rooms, at school, or even on their mobile devices. They use their online personas to express who they want to be or how they want to feel. When adults compare themselves to their Facebook friends' "ideal" profiles, their self-esteem may become affected. Because teens tend to place a lot more weight on how they are perceived by their peers, whenever they do not get enough likes—or even worse, when their friends share negative comments on their posts—they're even more susceptible to self-doubt and even depression. Aside from affecting their sense of self-worth, overidentifying with their online personas can also affect their friendships and romantic relationships. The consequence of this is a skewed perspective of how to navigate their romantic relationships and poor ability to resolve conflicts.

SEX, LIES, AND NEWS FEEDS

Most teens spend nearly all their time online. They're being introduced to new interests, new friends, and new ways to date and get sex. Social networks and dating apps are making it easier for teens to have sex more than ever.

Some teens I spoke to told me that their "hook-ups" (sexual encounters) begin with some sort of introduction to each other online. Many teens have sex, of course, but never before has it been so easy for them to express their need for sex—sometimes even without expectation for romance or intimacy. Even more than adults, some teens like sexting on their mobile devices, instant messaging on Facebook, sharing naked photos on Snapchat, flirting on Blendr, or soliciting each other on Tinder or Grindr.

Some girls send naked photos of themselves or bare themselves through "VC-ing" or video chatting. They do so believing that the photos won't be shared—"he promised." Many teens are also more forward about asking for sex. It's easier (and safer) to solicit sex from a girl behind a computer screen or cellphone than when you're right in front of her, and teens are getting used to this. In fact, many teens begin their relationship, whether sexually involved or not, through texting. Texting can be harmful because teens tend to feel safer sharing more information over a smartphone. On the other hand, texting can be beneficial to some teens because it allows them more time to get to know each other before deciding to meet someone in person.

Emma, 15

Chicago, Illinois

There was this guy I used to like in one of my classes. I didn't think he liked me 'cause it was like he ignored me most of the time. Then I got a text from him, so we texted for a while. He said he saw me around and that he liked me. It didn't go too far with him. He was boring. He didn't make me laugh. He was too pushy, and he was too full of himself.

🖒 Like   🗨 Comment   ➡ Share

I asked Emma if she and this guy ever talked on the phone. She gave me the "deer in the headlights" look, *What do you mean?* Needless to say, the way flirtation looks these days has drastically changed, and it continues to change at a rapid pace. Teens take this further by spending hours checking out each other's online profiles. Social-media profiles are where a lot of teens get their

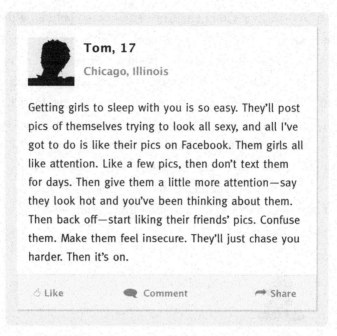

Tom, 17

Chicago, Illinois

Getting girls to sleep with you is so easy. They'll post pics of themselves trying to look all sexy, and all I've got to do is like their pics on Facebook. Them girls all like attention. Like a few pics, then don't text them for days. Then give them a little more attention—say they look hot and you've been thinking about them. Then back off—start liking their friends' pics. Confuse them. Make them feel insecure. They'll just chase you harder. Then it's on.

🖒 Like   💬 Comment   ➵ Share

information about their peers, and this new generation is deciding whether or not to hook up with someone based on what they find online. Scanning profiles makes it easier for them to get to know each other's interests without having to ask. Many teens I spoke to said they'll "stalk" a cute boy's profile and then stalk the girls the boy finds attractive.

When it comes to social media and dating for today's teens, it appears the more (online) information the better. What's alarming to me is that many teens seem to be basing some of their romantic/sexual decisions on what they're being presented through various online networks. They are perceiving interest through Facebook likes, competing with each other for attention and comparing themselves to each other. Many teens are also experiencing difficulty letting go of an ex. In the same manner as adults, they're tempted to continue to stalk their ex's movements post break-up.

Teens will follow their exes on several different networks, and when blocked some will even go so far as to stalk them through fake profiles.

Social media is causing many teens anxiety. Online interactions can be as confusing to them as they are for adults—if not more so. Not being followed on Twitter or liked on Facebook for many teens translates into rejection. Feeling rejected is painful. Instead of allowing themselves time to process or cope with their feelings in order to move on, many teens will move on in other artificial ways; they'll focus on someone else online.

MILLENNIALS AND THE DEATH OF PRIVACY

Unlike their adult counterparts, who grew up interacting with a significantly smaller group of friends (and defining friends as those met through real-life interactions), today's teens have almost unlimited options for their connections through the Internet and social media. The larger a teen's network, the more likely they are to have a wider variety of friends and share more personal information. Who are they friending and what are they sharing?

The Pew Report shows that the typical Facebook-using teen has 300 friends. Most teens are Facebook friends with:

- Peers from school (98 percent)

- Extended family (91 percent)

- Friends who go to different high schools (89 percent)

- Siblings (76 percent)

- Parents (70 percent)

- People they have never met in person (33 percent)

Dr. Rosen's study found that there are benefits to Facebook and social media when it comes to teen friendships. Social-networking can help introverted adolescents the same way it helps introverted adults—it helps them learn how to socialize behind the safety of their computer monitors. The study also shows that the more time teens spend on Facebook the better they are at showing "virtual empathy" to their online friends. This makes sense. Today's teens have grown up with the notion to like posts as a way to express support. Additionally, receiving news from their friends through News Feeds allows teens more opportunities to know what their friends are up to and how they are feeling. On the other hand, teens have also reported experiencing problems with their friends due to online interactions.

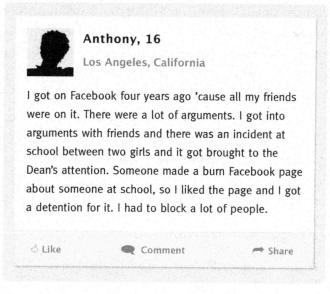

Anthony, 16

Los Angeles, California

I got on Facebook four years ago 'cause all my friends were on it. There were a lot of arguments. I got into arguments with friends and there was an incident at school between two girls and it got brought to the Dean's attention. Someone made a burn Facebook page about someone at school, so I liked the page and I got a detention for it. I had to block a lot of people.

🖒 Like   💬 Comment   ➡ Share

Typically, we learn socialization skills from our guardians or role models. We also learn them through trial and error when we interact with others. Ideally, we learn such skills through real-life interactions; however, many teens have grown up learning that it's okay to share and discuss anything online—even negative feelings about each other. Hearing your friend express negative feelings about you is tough enough, but reading them online adds a new dimension to the conflict. The public element adds pain, humiliation, and possible misunderstanding.

Patrick, 16

Chicago, Illinois

Over winter break this guy wrote on my girlfriend's wall. I got upset about what he said. He wrote that she was hot on one of her pics, and one of my friends commented on it. We got into it on her wall. Then he wrote on my wall that my girlfriend likes him. Then he brought more people into it. So all these people were fighting there. Then he started fighting with me on Twitter. My girlfriend wanted me to stop so I sent him an IM message telling him to lay off. He wrote back some shit so I just blocked him. Now he's trying to mess with me on Instagram.

🖒 Like   💬 Comment   ➧ Share

Patrick and his girlfriend broke up earlier this year shortly after his girlfriend switched schools. They are trying to remain friends, so they follow each other on Twitter, and every so often they will send instant messages (IMs) to each other on Facebook, but they do not speak to each other in person or on the phone. When I asked Patrick why he prefers to send IMs to his ex-girlfriend instead of calling, he smiled and responded, "We broke up. That would be weird." Another teen told me that she prefers to text and send IMs to her friends on Facebook instead of calling because she doesn't like the brief pauses of silence that occur between conversation topics when speaking on the phone.

Just as teens are communicating more online, it appears that they're also communicating less offline in terms of trying to resolve their disputes. Their real-life conversations are also taking a back seat to texting. As the example stated above illustrates, texting and communicating through IMs affords many teens instant access to each other, but it is also affects their comfort level with basic nuances of conversation. This shift in how teens communicate will inevitably affect their ability toward reflection, self-awareness and understanding of what should and should not be communicated.

Generation X is likely the last to know privacy as something to be guarded. The younger generation grew up with the social norm of self-expression through a public forum. This alone has altered their mindset about why and how they share information. In order to gain some sense of privacy, they will unfortunately have to learn the hard way; meaning, that through trial and error, teens will learn how privacy, or lack of privacy, on social media affects them.

One of my teen clients shared her understanding of Facebook privacy, "I haven't checked my privacy settings in forever...I don't think about that. I don't mind what people know about me." According to The Pew Report, sixty percent of teens on Facebook say they've checked their privacy settings in the past month and most believe that they are in control of their privacy settings. Some teens have made attempts at protecting their privacy while fourteen percent have profiles that are completely public. On the other hand, sixty-four percent of teens using Twitter choose a completely public profile and twelve percent are not sure which of their tweets are public and which are private. Some of this may be attributable to how social networks have set up default privacy settings, while other parts are due to teens simply not thinking about their privacy. What are teens sharing, where are they posting, and why are they sharing it? The Pew Report compared what teens shared in 2012 compared to what they shared in 2006. Overall, teens are sharing more information about themselves when it comes to photos of themselves, their school name, the city they live in, their email address and their cellphone number. The most significant increase involved them publicly sharing their cellphone numbers (up by 18%).

Privacy is taking on an entirely new meaning. Aside from sharing information about themselves, social media has enabled the use of location-based features such as checking-in and tagging. In fact, some networks are set up so that posts will automatically include location. While most teens are sharing more and more personal information online, there are some who are concerned about their privacy settings.

Lynne, 17

Austin, Texas

One of my teachers talked to us about it. She said that they [colleges] can check our Facebook accounts—even our old posts. I'm kind of freaked out. I went to a party and got really drunk. I looked drunk in a lot of the pics. I would've never shared them, but one of my friends did on her wall because she thought it was pretty funny. She tagged me on them. My parents saw that. I was so mad at her.

☃ Like   💬 Comment   ➞ Share

A study from the University of Illinois found that first-year college students are far from unconcerned about privacy matters and "the majority of young adult users of Facebook are engaged with managing their privacy settings on the site at least to some extent." As teens get older, it appears they better understand the importance of protecting their online privacy.

CYBERBULLYING

More than a million teens are subjected to extreme cyberbullying every day with Facebook accounting for more than half of the abuse. The main difference between bullying and cyberbullying is that when teens get bullied online, hundreds, if not thousands of people, experience and take part in the harassment. For this reason, cyberbullying can go on for months. One single post, on any social-media site, can trigger an onslaught of other harassing posts.

Lana, 16

Makawao, Maui, Hawaii

I was Facebook chatting with one of my friends when we got into an argument. My friend started calling me 'fat' on her wall. Ever since, I worried about my weight, which escalated to anorexia and bulimia. I struggled with this for a couple of years—all due to this one interaction. I eventually got off of Facebook, and just used Tumblr and Vine instead.

Like   Comment   Share

Some think the inherent distance of social media causes teens to post bolder and more hurtful comments than they would ever say to each other face-to-face. Other teens may then feel free to weigh-in or add to the harassment even when they know nothing about the situation.

One story in particular caught my attention and forever changed the way I perceived the effects of cyberbullying. I often present this story during my workshops to students, teachers and parents at local high schools and colleges—the story of Amanda Todd:

> On September 7, 2012, teen Amanda Todd posted a nine-minute YouTube video entitled, *My Story: Struggling, Bullying, Suicide, and Self-Harm*, which showed her presenting a series of flash cards relating her experiences being bullied. According to Wikipedia, her video went viral, receiving over 1,600,000 views by October 13, 2012.
>
> During her video, Amanda writes about her experiences in 7th grade where she video chatted with a stranger who convinced her to bare her breasts on camera. He then blackmailed her by threatening to expose the topless photo to her friends unless she exposed herself even more. This man followed her through Facebook over the next two years. Police informed her family that Amanda's photo was circulating over the Internet.
>
> On the YouTube video, Amanda wrote that during this time she began to experience severe anxiety, depression and panic attacks. She began using drugs and alcohol. The stranger continued to harass her and created a Facebook profile and used the photo of Amanda's naked breasts as his profile picture. He contacted her classmates at her new school leading her to again be teased and bullied. Amanda

was forced to change schools a second time. Following a physical attack at her new school, Amanda made a suicide attempt by drinking bleach, but survived after being rushed to the hospital. She was later harassed and teased online by her peers for not "properly" committing suicide. Her family was forced to relocate her again.

Amanda was unable to escape her past. Every time she moved, the stranger that originally blackmailed her would find her on Facebook. He tricked her into friending him under an alias and would again send Amanda's photo and video to her new school, peers, teachers, and parents. Amanda's psychological state worsened and she began to cut herself on her arms. She again attempted suicide and continued to be harassed by her peers. On October 10, 2012 Amanda was found hanged in her home.

Stories like Amanda's are heartbreaking to read. She continues to be harassed, even after her death, on "burn" pages on certain media sites. This shows the magnitude of cyberbullying and the extremely harmful effects it can have on a teen.

Some parents and school administrators may believe that cyberbullying is not a true form of bullying because they believe teens can easily delete a comment, block a friend, or deactivate their social-media accounts and the problems disappear. This isn't the case. Once something gets posted it can go public very quickly. With so many people seeing the content, teens can become upset, nervous, or frantic about the social effects of the post.

ONCE THE BULLYING BEGINS

Teens may not know to whom to turn. If the bullying gets serious enough they will report the harassment to school officials or the police. Unfortunately, many school officials do not know how to address cyberbullying.

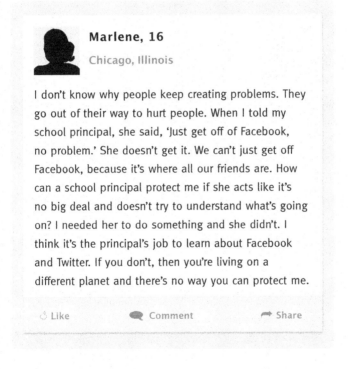

Marlene, 16

Chicago, Illinois

I don't know why people keep creating problems. They go out of their way to hurt people. When I told my school principal, she said, 'Just get off of Facebook, no problem.' She doesn't get it. We can't just get off Facebook, because it's where all our friends are. How can a school principal protect me if she acts like it's no big deal and doesn't try to understand what's going on? I needed her to do something and she didn't. I think it's the principal's job to learn about Facebook and Twitter. If you don't, then you're living on a different planet and there's no way you can protect me.

☺ Like   💬 Comment   ↪ Share

Marlene is absolutely right; school administrators must make efforts to understand their students' interactive world. How else can they conceptualize what cyberbullying is or how it emotionally affects teens?

Liam Hackett, founder of the national antibullying charity, *Ditch The Label*, which produced the largest survey to date on online abuse, said that many people assume cyberbullying is not as hurtful as face-to-face abuse; however, it can be even more distressing because of its public element. Cyberbullying can be catastrophic to a teen's self-esteem.

Tessa, 16

Chicago, Illinois

I was a cheerleader my sophomore year and we went to camp. I was roommates with this girl. I was kind of scared of her because she would take advantage of me and use me, so I asked to not be in the same camp room as her. Everything went well until the last day. We were all sitting in the hallway when she tripped over my feet and her towel fell off. She thought that me and three other girls pulled off her towel, but we didn't. The next day my mom was on her way to have a meeting with the school about this. Apparently her mom went to the cops and wanted to file a battery and assault but nothing was charged. From that moment on, I've gotten Facebook messages and texts saying, 'Your mom can't save you forever,' 'that little cunt called me out,' 'you're like a disease—you keep coming back,' 'Tessa's a bitch.' I've had to change my phone number like six times. This has been going on for the last year and a half.

      Like   Comment   Share

A recent cyberbullying survey found more than one-million teenagers suffer "extreme" abuse online daily and that levels of cyberbullying are much higher than reported. The rise of cyber-bullying has called for parents and regulators to recognize the seriousness of the issue.

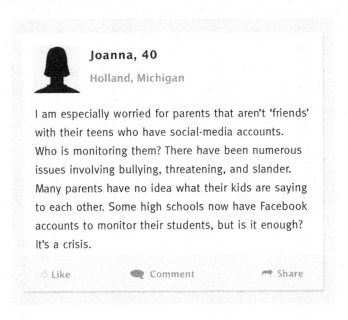

Joanna, 40

Holland, Michigan

I am especially worried for parents that aren't 'friends' with their teens who have social-media accounts. Who is monitoring them? There have been numerous issues involving bullying, threatening, and slander. Many parents have no idea what their kids are saying to each other. Some high schools now have Facebook accounts to monitor their students, but is it enough? It's a crisis.

Like   Comment   Share

WHAT CAN WE DO AS PARENTS?

As teens continue to engage in online interactions, we need to make greater efforts to educate teens, parents, and school officials about the dangers of cyberbullying. Many social-media sites, including Facebook, have strict policies in place to address cyberbullying, but it may not be enough. I believe we also need stricter governmental laws in place to protect teens. For now, there are certain steps parents and school officials can take to ensure that teens are using social media appropriately:

1. *Study social media*: Learn about as many social-media networks as you can. Make efforts to understand the networks and how your teen interacts with each of them. Ask your teen questions, do online research into each social-media site, and study the current trends. There is really no way around this first step. If you want to understand your teen you must begin by examining the way they connect, engage and communicate with their peers. Studying various social-media sites will be time consuming, but in order to help them you have to be able to relate to them.

2. *Don't try to control all of your teen's social-media networks*: Instead talk with your teen about how to appropriately use social media. Encourage discussions to talk about what you see on these social sites. Ask your teen what they think about the different sites, which they prefer and why. Encourage direct communication about problems that they face online. Explain to them why privacy is important and how to determine what they should share on social media, taking time to explore what they should keep to themselves. Remind them that social media was created to complement real life, a way to deepen their real-life connections, not separate them. Finally, open up a conversation about cyberbullying.

3. *Attend to your teen's privacy settings*: If you don't know how to address privacy settings on your teen's social-media sites, ask others for advice. Privacy as we once knew it no longer exists. Parents, teachers and school officials, and law enforcement need to accept and adapt to this reality. Teens are tech savvy but they're not taking the time to question

how the lack of privacy affects them. Sit with them and sort through the challenges together in an active, fun, and constructive manner. Ask your child to teach you since they're so adept at using social media. For further advice, consult with Information Technology (IT) or social-media specialists who can provide you with the most current instructions on how to set stricter privacy settings for your teens. The nature of social media will make it virtually impossible to give you and your teen the privacy you want, but do what you can to keep identifying information as private. Navigating this process with them, not for them, is the first step to understanding and protecting how they use social media.

4. *Speak to school about their strategies for addressing online issues*: Become an informed parent and find out what your child's school has in place to protect students. If your teen's school does not have any strategies in place, ask them why and when they plan to implement them. Become an active advocate to get such policies in place as quickly as possible. Organize discussions between parents, teachers and school administrators on how social media is affecting today's teens and explore various ways that adults can protect them.

5. *Report Cyberbullying*: When cyberbullying happens, it is important to document and report the behavior to schools, online service providers, and law enforcement if needed. Don't forward or respond to threatening or intrusive messages. Instead, keep evidence, including dates, times and descriptions of cyberbullying. Save and print relevant screenshots, emails, and text messages.

Cyberbullying often violates the terms of service established by social-media sites and Internet Service Providers. Therefore, it's important to review their terms and conditions or rights and responsibilities sections. These describe content that is or is not appropriate. Report cyberbullying to the social-media site so they can take action against users abusing the terms of service. Be sure to visit social-media safety centers to learn how to block users and change settings to control who can contact you.

When cyberbullying involves harassment, stalking behavior, hate crimes, threatening statements, child pornography, sexually explicit messages or photographs, or photographs or video taken where a person expected privacy, it is considered a crime in many states and should be reported to law enforcement. Some states consider other forms of cyberbullying criminal. Study your state's laws and law enforcement for guidance. This may all seem like a lot of work, but taking a stand may set an example that can prevent thousands of children and teens from getting hurt within your community as well as other communities.

WHERE DO WE GO FROM HERE?

There is no doubt that social media is affecting how teens process information, perceive themselves and others, resolve conflicts, and cope with challenges. Parents are often confused about how to address social media. Try to accept social media as an integral part of today's teen generation, and take steps to understand this new world and your teen's place in it. This should make you better able to help your teen avoid social-media-related drama. Make spending quality time communicating and interacting with your teen a priority.

Allow and encourage your teen's appropriate self-expression online, but limit the amount of time they spend logged on. Think of activities that you and your teen can do together. Take your teen fishing, camping, on a family road trip, or even just on a long walk or drive, so you can get a chance to talk without the distraction of electronics. Meet his or her friends by inviting them over for dinner and movie night. Highjack their smartphones for a couple of hours if necessary. Do whatever you have to do, but take the time to engage with your teen.

In today's busy world, parents often feel overwhelmed with multiple obligations—I get it. However, the less you interact with them, the more they will turn to others for advice and support. Ultimately, no matter how many times teens may protest, they still need the structure, protection and support that only a parent or guardian can provide. The more you show interest in understanding them and their world, the stronger your relationship with your teen will be.

chapter seven

With Friends Like These, Who Needs Enemies?

After a while I didn't recognize myself anymore. I got sucked into his craziness. His Facebook stalking became my Facebook stalking. His jealousy became my jealousy. It was all a game. He was trying to see if he could break me and, for a while there, I thought he did.

THE DARK SIDE OF FACEBOOK: EMOTIONAL MANIPULATORS AND PSYCHOLOGICAL DAMAGE

ONE OF THE DARKER SIDES OF FACEBOOK INTERACTION IS when we find ourselves connecting with Emotional Manipulators. Whether they are people we know personally, or people we meet online, these are individuals who (at first glance) appear friendly, intelligent, funny, noble, and very charming. They spark interest in others through their elaborate and creative posts. When we communicate with them directly they "connect" with us in a very profound way.

Emotional Manipulators make us feel comfortable sharing our personal goals and dreams. We tend to trust them easily. After a while, however, the comfort we initially felt when we met them begins to subside and turns to confusion, irritability, anxiety and even fear.

Emotional Manipulators make it a point to be perceived favorably through using many underhanded or deceptive tactics. Initially, they will spoil you, adore you, charm you, praise you, and make you believe that you've found a Prince Charming or a Greek Goddess. They make you the center of their universe and provide you with endless support and encouragement, all with one

motive in mind: to emotionally control you. If they can successfully affect your self-esteem or alter your values or beliefs, they feel
that they've "won." On Facebook, toxic personality types can take

Tanya, 41

Charlotte, North Carolina

Brian sent me a friend request after finding me on a
mutual friend's list. I consider myself attractive and I
have many friends. I noticed that Brian was intelligent,
worldly, cultured, well-traveled, and exceedingly
supportive of all his friends. Brian began private chats
with me and told me he'd recently broken up with his
fiancée after she cheated on him. I felt compassionate
toward him and wondered why any woman would
ever cheat on such a great guy. He had it all: good
looks, charm, generosity, and sensitivity. After some
time, however, his comments on my page began to
slowly change. He provided a mixture of supportive
comments followed by subtle criticisms. And there
were inconsistencies in his stories: First he said he
worked at a pharmaceutical company, then said he was
teaching. He initially said he never married, but later
mentioned his ex-wife. When I confronted him he told
me I was confused and was 'too sensitive.' Whenever
I caught him in a lie, he convinced me that he had
good intentions and that I was the one sabotaging our
relationship. No matter what Brian did, it seemed that I
had to take the blame.

🖒 Like   💬 Comment   ➡ Share

on several forms. The most common types are: the Saboteur, the Narcissist, the Martyr, the Seducer, and the Stalker.

The Saboteur

A Saboteur is someone who you believe is your friend, sometimes acts like your friend, but secretly resents you. Saboteurs are masters at discovering a person's vulnerabilities and then using them to their advantage. At the beginning of any relationship, they are generous with praise and flattery in order to hook their emotional victims. When they believe they've gained your trust, they slowly and systematically begin to alter your self-esteem through subtle insults and criticism, which they will describe as friendly suggestions "to help you grow" as a person. If you protest their criticisms, they will accuse you of being too sensitive.

Saboteurs tend to be self-centered and have difficulty maintaining healthy, long-term friendships. They like to provoke emotional reactions. Often motivated by jealousy of other people's success and happiness, they create mayhem and sabotage their friends, playing people against each other, spreading rumors and planting seeds of doubt and fear in someone's mind. Even if they can't stand a person, they will check that person's Facebook wall almost constantly. When "crossed," they often feel the need to get revenge, even if there's no legitimate reason for their actions.

Saboteurs want to ruin reputations; they emotionally attack others but never want to be "the bad guy." They blame other people or circumstances for their actions. They may be seen posting jabs at their Facebook friends and then quickly removing them from their wall. When confronted, they reply with something like, *See I knew you'd over-react and think that was about you!* They are masters at

Linda, 30
Seattle, Washington

Steven said that he would always be there for me. I was feeling especially vulnerable when we met because my family didn't support my career choice and our relationship was strained. Steven was definitely there for me, offering support and encouragement. But he didn't want me to tell my family about him, saying, 'They won't understand and will turn you against me.' Oddly enough, he also encouraged me to not trust my other friends. One day, Steven got angry when I didn't agree with his advice. He started calling me 'stupid' and 'crazy.' I couldn't understand why he was acting this way. When I told him how hurtful he was being, he'd respond, 'Well, then just go back to your crazy family—they seem to know what's best for you.' When it came to friendship with Steven, you were either with him or against him—there was no middle ground.

🖒 Like   💬 Comment   ➡ Share

projecting their feelings onto you.

Psychologist Melanie Klein described the process of *Projective Identification*—where one person subconsciously projects aspects of their own personality onto another person. The person on the receiving end of the projection often experiences a loss in insight or identity as they get caught up in and manipulated by the other person's fantasy. For example, when someone is threatened by you, unconsciously they may need to perceive you as vulnerable or susceptible and in need of their protection. They will then take steps to try to dominate you—in some way. The drive behind this is that they themselves feel emotionally vulnerable and fear being emotionally dominated by you. Saboteurs use the following tactics to manipulate:

- Appearing trustworthy when they are not

- Forming false alliances with others in order to hurt you

- Blaming you for their actions

- Planting negative ideas into your mind or the minds of others

- Telling multiple lies and exaggerating on their walls

The Saboteur's intentions seem noble at first and so you tend to place a lot of trust in their decisions. They convince you that they only have your best interests in mind. Sooner or later, however, you slowly begin to realize that their "good intentions" were really a mask for them getting their own way. And when you begin to push back, you will be met with resistance and even more projection—they will call you insecure, paranoid, mean, or crazy. Friendships or relationships with Saboteurs do not end well and often you will be blamed for their behavior because many of them cannot tolerate even perceived criticism.

Once you realize that you are dealing with a Saboteur, when possible, it is best to keep them at a distance or not be friends with them at all. Ideally, it's best to have as little contact with them as possible. Try to remove workplace Saboteurs from your friend list or change your privacy settings, so they can't see your new posts. Another option is to announce that you have decided to keep "work at work" and therefore will be removing workplace colleagues from your friend list.

The Narcissist

A Narcissist is completely self-absorbed. This doesn't necessarily mean they are selfish. They often do wonderful and kind things for others. They often post about the many charitable events they attended or the honorable acts they have done for their fellow man. The main difference between those who altruistically help others and a Narcissist is that the Narcissist helps others to receive praise from as many people as possible. Without praise, their actions are pointless. These are people who post about their charitable acts in order to receive validation and praise for their good deeds.

A Narcissist is especially vulnerable to criticism, even if it's only in their head. If they feel intimidated by your post, they will quickly try to one-up you. Narcissists analyze all Facebook interactions for how the posts concern them. The two primary questions Narcissists also ask themselves when they interact on Facebook are: What can be done for me? How am I being perceived? Everything is about them. If you add a post to your own wall about a purchase you made, they will quickly point out their "better" purchase. If you comment about your son's spelling bee award, they will praise your son while "casually" noting that their son is on the honor roll. Any post, any like, or any comment contributed by a Narcissist is created in order to draw them more praise.

Narcissists have a strong need to feel superior toward practically everyone. You can never be "better" than them. They need to prove you wrong in order for them to be right and enjoy intimidating everyone around them. On Facebook, they add elaborate, almost theatrical, attention-seeking posts. You'll see multiple profile pics (sometimes over 50 photos) showing off their best features. The more endorsements and "likes," the better. They can be overwhelming.

Narcissists have a strong desire to be the best of the best. They may use intimidation to get the praise they believe they deserve. While some display obvious arrogance, many are subtle in their

Jennifer, 40

Washington D.C.

Emily's three girls spent many hours at my pool playing with my three boys. Our families became very close. I friended her on Facebook and, at first, things seemed normal enough. Just like any mom, Emily posted updates of her children and other family-related events. But then, Emily's behavior on Facebook changed. On certain days Emily was friendly while on other days she wouldn't even acknowledge my presence. I noticed a pattern: if I didn't 'like' Emily's status or photos, Emily was nasty toward me and my family. But if I liked them, she was fine. Emily began posting passive-aggressive vague posts, which I believed were directed toward me. One day, I noticed her status about removing 'negativity' in her life and she then proceeded to block me and my husband. And she told her children to ignore mine. In an effort to get to the bottom of this, I called her. Emily said, 'Well, you never like my posts! You just ignore them!' When I denied this, she replied, 'No, you liked the video of my daughter on my cousin's wall—not mine!' I was shocked and decided that it was best to not be friends anymore. I later discovered that Emily continued to stalk my posts through a mutual friend's wall.

🖒 Like   💬 Comment   ➥ Share

grandiosity, and these subtle Narcissists tend to be the best at manipulation. If they throw out a jab at you on Facebook, they'll add that they're "joking." If you return a jab, they will criticize you, blame you for their comment, and then insult you for not being able to take a joke. At the heart of a narcissistic personality is a very fragile ego and a need to hide shame. They do everything they can to convince others that they're flawless.

Everything the Narcissist says or does is about status. They can be very materialistic and simply don't understand that others don't share their values. In relationships, they are especially hurtful. They expect everything from their partners and will emotionally attack if their partner dares refuse them or criticize them. Narcissists use Facebook as a platform to insult or hurt their partner. They use the following tactics to manipulate:

- Initially being exceedingly charming, funny and complimentary

- Posting group photos, but only those that flatter them

- Adding elaborate posts filled with details of their experiences in an effort to solicit positive commentary

- Filtering comments by deleting negative ones

- Posting only things that make them look good

Most Facebook Narcissists are not as intrusive to your online life as they can be in real life, but their posts may still be draining. Simply hiding them from your News Feed will spare you from their posts. If they criticize your posts, set boundaries with them by telling them how their comments make you feel. If this problem continues, remove them from your friend list with no regrets.

THE MARTYR

Martyrs tend to be on Facebook constantly. Their posts frequently portray them as victims. They might say, *I'm so fed up with people. No one cares anymore. I'm just going to give up and go away!* to trigger friends to respond with words of encouragement and compassion.

Nicky, 34

Milwaukee, Wisconsin

Ann and I were friends in high school. When I found her on Facebook, we had many mutual friends in common from the old neighborhood. We quickly became friends again because we have a lot in common: We were both career driven, love our pets, love to run, and share the same tastes in books. But Ann seemed to always fall on hard times. Whether it was financial hardships, medical problems or feeling misunderstood by others, she just couldn't seem to get it together. I helped her financially and supported her emotionally, but it was never enough. It was getting to be too much. No matter how much encouragement or advice I gave her, she ignored it, and seemed more interested in focusing on the negative. I suggested she see a therapist, but she laughed it off. 'Nothing will help me,' she told me. I wanted to be there for her but after a year of this, I'd had enough. When Ann noticed me becoming distant, she posted a single sentence on my Facebook wall: 'You're just like all the others.'

    ☺ Like   💬 Comment   ➞ Share

There is nothing wrong with seeking comfort from friends and loved ones on a tough day. What makes the Martyr a toxic presence on Facebook is that even at the start of a friend connection, they appear to be vulnerable. For example, they may carefully observe which events you are attending and comment on your photographs with a comment such as, *I wish I was invited; I would've loved to wish you well too! (just kidding!)* Guilt is a powerful weapon used by some Martyr personalities.

Unlike individuals who strive to overcome past or current challenges, Martyrs have a strong need to act victimized. They surround themselves with people who will show them compassion and feed into their helplessness. Martyrs have often grown up believing that they deserve special treatment and continue to expect it in adulthood. They identify others' vulnerabilities and manipulate them to get support. Many don't possess the skills to manage life, but they'd rather you manage their life for them than try to improve their circumstances themselves.

Martyrs have a difficult time seeing another person's point of view and can be self-centered and demanding. At first, they motivate you to help them through flattery and gratitude. But if you avoid them, they may rapidly become angry and nasty toward you. They have a strong need for support and attention. Martyrs use the following tactics to manipulate:

- Using guilt trips to get their way

- Posting about their disappointment when feeling ignored or left out

- Playing on their friends' compassion

♀ Sharing many perceived hardships on their wall

♀ Hinting at their helplessness and hopelessness for all to see

Many Martyrs on Facebook are our family members or close friends in real life, which makes coping with their posts especially challenging. However, unfriending them on Facebook may cause them to lash out, leading to dire social consequences. When it comes to coping with these emotionally challenging individuals, it may be best to simply hide them from your News Feed. You can still check on them from time to time, but hiding them from your News Feed will help you feel less overwhelmed by their postings.

Martyrs need attention, but they can't get all the attention they seek from one person. Setting boundaries is often necessary with Martyrs. Encourage their involvement in community or support groups that can better provide them with the kind of support they are seeking. Explain to them that as much as you would like to help them, you can't do it alone. What they ask for and what they need are two very different things; when we're able to set limits with someone, sometimes we empower them to feel capable of coping with situations on their own.

THE SEDUCER

Seducers on Facebook tend to be somewhat superficial and believe themselves to be physically attractive. They almost constantly post photos of themselves. In real life, some are very sexually active or sexually tease to get what they want. On Facebook, they flirt with various people at the same time and test people—even married

Frank, 46

Portland, Maine

My father had died when I met Debbie on Facebook. She said her grandfather had just died too, and I guess that was one of the things that connected us. My wife and I were having problems, so it was a bad time for me. Debbie and I started out as friends. I didn't mean for anything to happen, but she just got me, you know? She understood everything about me that my wife didn't, and she seemed amazing. She posted pics of her at charity events and doing all this important stuff. When Debbie and I met, it was amazing and the sex was even more amazing. But it wasn't just the sex. I really fell in love and wanted to spend the rest of my life with her. I asked my wife for a divorce, and I lost my kids—she took them to Vegas and, like a moron, I was dumb enough to give up custody. I regret that, but back then all I could do was think about Debbie. It was like I was brainwashed. When I was free to begin my life with Debbie, she told me she was confused and needed space 'cause everything was moving so fast. I didn't know what she was meant. She disappeared and didn't talk to me. Then on Facebook I saw that she was flirting with one of my buddies. I lost everything over this bitch—my wife, my kids, and my job 'cause I couldn't focus at work. I tried to warn my buddy about her, but he said she's like no one he's ever met before. I cut him off too. He's going to learn the hard way. They're not my problem, right? I hate her so much, I want to warn everyone about her.

👍 Like   💬 Comment   ➦ Share

people—to see if they will flirt back. They have an intense need for sexual and romantic attention, and tend to be very competitive with spouses of people they are interested in. There are many reasons why they use flirting and seduction as a game. Some do this due to having deep-seated insecurities, others have been wounded romantically or are acting out past traumas.

Seducers use the Relationship Status feature to their advantage: It's Complicated. Whether or not this is actually true, this relationship status is usually a manipulative move. When probed for more information, they use lines such as, *I have a boyfriend but we're having problems. He just doesn't understand me.* What follows is a compassionate response from those who will listen. Seducers often proclaim that they are misunderstood and victimized by their significant other.

Being the favored child typically forms the Seducer's personality. They have grown up with a lot of attention and need the attention to continue into their adulthood. There is also evidence that some Seducers may have experienced a history of sexual abuse. They have learned that in order to get and maintain attention, they have to play with people's thoughts.

The Seducer's interactions are considered by many to be inappropriate and dramatic. At their most toxic level, Seducers feel most powerful when they can destroy a person or a family. Their aim is not involvement in a healthy romantic relationship, but to seduce. Their victims are often in an emotionally vulnerable state leaving them susceptible to suggestions and false promises.

Flirtation is a system of behaviors rooted within each of us. People like people who like them, but people also like people who might like them. This intrigue is what Facebook Seducers use to seduce their victims. You can never be completely sure about their feelings. You want them to like you and only you, but very often what they say and what they actually do are two very different things. Uncertainty is the Seducer's greatest weapon.

A recent study conducted on 47 female undergraduates tested how attracted we are to uncertainty. Each woman was told that several male students had viewed her Facebook profile and rated how much he'd like to get to know her. Women were most attracted when they were led to believe that the men were "uncertain" about them. As it turns out, being unavailable isn't attractive but being mysterious is. Furthermore, it wasn't necessarily the uncertainty that attracted the women, but the thoughts they were having about these men as a result of their uncertainty.

It's easy to become fixated on a Seducer when uncertain of their interest, and mysterious or vague wall posts only lead to greater confusion and frustration. The target may confuse the thoughts they're having about the Seducer to mean that they're in love, allowing the Seducer to say and do almost anything they choose. Still, their victim will remain hooked. Seducers use the following tactics to manipulate:

- Contacting you and appearing vulnerable

- Designating you as the "special" one in their life

- Displaying provocative behavior that is not appropriate to the situation, such as flirting with a married woman

- Convincing you that something is wrong with you when confronted

- Leaving the relationship abruptly when they've "conquered" their victim

Because their interactions often cause one to fall in love or become infatuated with them, Facebook Seducers tend to cause a lot of emotional damage. They can have a crushing effect on a person, either through purposeful manipulations or passive flirtations. Seducers also typically have a difficult time respecting boundaries; their online and offline interactions can earn them court-ordered restraining orders. For this reason, firm limit setting, unfriending, or blocking them from your account is often required. Benefit may be had from taking a Facebook break if you have experienced a Seducer until you are able to see the interaction a bit more clearly.

The Stalker

The Stalker loves to snoop and get as much information as they can about you and your interactions. Facebook Stalkers get a lot of pleasure from watching a situation develop from an intimate exchange between two or more people. Facebook almost encourages stalking because one can snoop into multiple people's lives at the same time. However, the Facebook Stalker doesn't feel like a snoop at all. After all, it's not really snooping if personal information is willingly offered up for all to see.

Facebook stalking is more about enjoying the intimacy of the situation, similar to being a "fly on the wall" to a conversation that is typically a highly personal exchange between two people. Being a passive observer to this intimacy can be highly arousing—in a nonsexual way—for some people. They relish the information they receive; in a sense, they claim to own it. They view your personal life as their personal property. Once they have it, they can do with it as they please.

Reese, 19

Helena, Montana

My boyfriend ended our relationship after six years—just like that, without warning. I kept texting him but he didn't text back. I called him but he wouldn't pick up. He never even told me why he broke up with me. I started stalking his Facebook page for something I could find that would make sense. When he friended a new girl, I checked out her page and tried to find anything he posted on her wall. I needed to know who he was dating 'cause I wanted to talk to her. I finally figured it out and sent her some messages. I tried to warn her what he's really like and told her that he would do the same thing to her. I wasn't stalking her—I was trying to help her, but she wouldn't listen. She told her parents who ended up calling the police. The cops came to my house and spoke to my parents. It was awful. I promised I would leave them alone. I don't contact them anymore, but I still follow my ex—I just use an alias so he won't find out it's me.

    🖒 Like   💬 Comment   ↪ Share

Research by University of Missouri Professor Kevin Wise showed that healthy wall viewing in Facebook consists of what he terms "social browsing," where friends and family look at your general News Feed and updates, enjoy the read but then move on to other people's News Feeds. "Social searching" involves a more intense motive from the viewer. When someone "social searches" they focus merely on your wall. This viewer is acting as if he or she is obsessed with you.

Stalking on Facebook can go beyond eavesdropping. They get to know about you, ALL about you. They don't respect privacy or boundaries. What may start out as harmless curiosity for these people can lead to what stalkers refer to as "Facebook research" and even compulsive stalking.

Most people understand what constitutes a reasonable amount of Facebook browsing in the early stages of a relationship. If you've been asked out on a date, you "do a little research" if you're

Louise, 26

New Orleans, Louisiana

I have a Facebook stalker. He's a guy I used to date years ago, but since I've been with Joe and changed my Facebook status to engaged, he won't leave me alone. This fool called my phone when he knew Joe and I would be home. It was like he was intentionally trying to cause a problem in my relationship. I blocked the calls from then on. Now instead of calling, he sends me endless messages on Facebook especially when he sees that I'm online making a status update or liking someone else's comment or picture. On my wall, he asks if I miss him and if I want to go out one day. I have to keep deleting his nonsense. I hope he'll get the hint but he hasn't yet. Men seem to want what they can't have and the more unattainable the conquest is, the harder they push for it. I wonder how long it will take for him to stop sending me messages.

☺ Like   💬 Comment   ➞ Share

interested. We want to know more about whom we're interacting with; this is normal. It's also tempting to repeatedly check an ex's page post break-up. In fact, according to a statistic on social-media-marketing site Socialpeel.com, more than 60% of Facebook users check their ex's page. If we're Facebook friends with someone it's understood that you have access to their posts. But how do you know when checking statuses goes too far?

Facebook Stalkers have a hard time leaving you alone. They enjoy searching your wall a bit too much. They may repeatedly post on your wall or send you multiple messages. While they are not necessarily being mean or threatening, the behavior becomes intrusive and, at worst, scary. Stalkers manipulate through these tactics:

- Going through all your photo albums and wall posts in order to obtain as much information as possible

- Friending your Facebook friends in an effort to remain connected to you

- Creating false Facebook accounts when blocked from your account

- Leaving you messages on other people's Facebook walls

- Continuing to send you messages despite repeated subtle or overt messages to leave you alone

Not unlike Facebook Seducers, Stalkers also have a difficult time honoring a person's boundaries. When it comes to addressing your concerns with them, stricter limits may be necessary. When blocked from someone's profile they are known to create false Facebook profiles to continue stalking behavior. Needless to say their understanding of what is appropriate versus what is not is very distorted.

When they do not respond to silence or limit setting, a Stalker's actions often cause many Facebook users to take drastic measures in order to remove them from their life. No one has the right to cause you undue anxiety and fear. If someone refuses to leave you alone you may need to directly contact Facebook or local police for assistance.

Jackie, 29

San Francisco, California

I believed I had found the man of my dreams in Seattle. According to his profile, Justin was professionally successful, believed in long-term commitment, and wanted to have a wife and children one day. I began communicating with Justin when he friended me through a mutual friend. I instantly fell in love and even posted on my wall that I was very happy and was considering moving to Seattle. I bought a ticket to Seattle, but before I left I decided to ask some mutual friends more about him. I discovered that Justin had two children with different women, did not provide child support to either of them, and never even graduated college. I felt completely betrayed and heartbroken. I was so depressed that I couldn't even get out of bed. Just thinking about it made me have panic attacks. It got so bad I had to go to therapy. When I finally unfriended him, he proceeded to publicly slander me on his own wall.

ô Like   🗩 Comment   📣 Share

THE PSYCHOLOGICAL IMPACT OF EMOTIONAL MANIPULATORS

Whether for sport or for compensation, false representation on Facebook is mostly harmless, but when sport becomes malicious behavior, people can end up being hurt or traumatized.

When we begin to realize we are being manipulated online and see someone for who they really are, we feel confused and angry, as Jackie did. Confusion and anger can give way to anxiety and eventually self-doubt: *Why would someone act this way? Was it me? What made me fall for this?* If the growing feeling of self-doubt persists, what happens next is the most psychologically damaging: many will spend hours, or days, overanalyzing, interpreting and misinterpreting all of these comments and posts, and this becomes a way of living, a way of defending and having to protect oneself. We search through our comment history, trying to make some sense of how this could have happened, without understanding that it's not possible to find the "real" meaning or logic to this behavior, because there is no real meaning or logic. People post what they want to post. Our reaction—whatever it may be—gives Emotional Manipulators the ultimate payoff they were hoping for. For some, even negative attention is validation because it's still attention.

Interactions with emotionally manipulative personality types can seriously affect your emotional well-being. Through your interactions with them you may become fearful and depressed, but feel that you cannot let them go. You may notice that you apologize a lot to them or that you've changed your own behavior around them. You've probably also grown accustomed to making excuses for them or tried to help them because, on a certain level, they've convinced you that their behavior is your fault.

Where Do We Go From Here?

The first step in beginning to heal from interaction with an Emotional Manipulator is to realize that the Manipulator knows exactly what he's doing. They fully understand the effect their Facebook posts have on your life. In fact, they're counting on it. Whether the Facebook Manipulator is a Saboteur, a Narcissist, a Martyr, a Seducer, or a Stalker, their aim is to get what they want without any consideration to your feelings. Once you realize that you are interacting with an Emotional Manipulator, you can take steps to remove them from your life.

Start by noticing the difference between what the Emotional Manipulator posts on Facebook and what they actually do. Emotional Manipulators almost always start out as charming and captivating individuals. They make you the center of their universe. You may feel like you're getting everything that you've ever wanted in a partner/friend, but the affection they display is given a bit earlier in the relationship than is appropriate. They make multiple promises and "talk a good game" on their Facebook walls, but with some investigation, you may notice inconsistencies in their presentation.

Write a list of how they present themselves versus how they actually treat you. Emotional Manipulators often have grandiose plans that never come to fruition. They are fantasy-driven; their goals are a bit far fetched.

Write a list of inconsistencies they have either posted or shared with you personally. Ask yourself if they're still lying or if they've shared conflicting stories. Consider whether they've given you any false hopes or if their attention and support was conditional. Writing down the discrepancies will help you get a clearer picture of what is going on and will make it easier to distance yourself from an Emotional Manipulator. Resistance to manipulation can begin only when you understand the manipulator's intentions and methods.

Be aware of stalking behavior. Do they fail to leave you alone despite numerous hints and attempts to stop leaving wall comments or sending you messages? Are they trying to make contact through mutual friends? Are they suggesting that your relationship is closer than it actually is? Are you noticing a lot of commentary on photos of your partner or family members? Are they leaving sexually suggestive or inappropriate comments?

If you notice any of these signs, it may be time to firmly set limits and block them from your profile. If you believe that they've created a false account or have tried to contact you through mutual friends, inform your mutual friends of the situation and ask them not to share any information about you and not forward you any messages from them. If the situation continues, report them to Facebook or contact local authorities. Being stalked on Facebook is not the same as being stalked in person, but the emotional effect is the same. There is no reason why you should have to put up with inappropriate or abusive behavior.

A clear sign of an Emotional Manipulator is that they will often try to isolate you from those you trust. They don't want anyone else interfering in their plans. Let your friends know what you're experiencing. You may find that all you need is support and an objective point of view on the situation. When it comes to setting boundaries with an Emotional Manipulator, the more support you can get the better. Increase your communication with your other friends and use silence against the Manipulator.

When you're involved in a toxic relationship for a prolonged period of time, you may find yourself adjusting to the dysfunction.

You may not realize how unhealthy a relationship is until you've left it. Once you've stopped communication with the Manipulator and increased communication with your healthier friendships, you will begin to feel relieved that they are gone.

Emotional Manipulators have a lot of wants, needs and demands and they assume that you are in their life to give them what they want or to turn a blind eye to their behavior. Whether they demand attention, money or praise, giving an Emotional Manipulator what they want is not always what they need. Focus on what is a healthy versus an unhealthy friendship. If you're feeling overwhelmed, put down, insulted, confused, or emotionally attacked you are likely involved in a toxic relationship. Even on Facebook, comments carry a punch and even the best of us get punched sometimes. A Manipulator will try to get a reaction, by any means possible, but that doesn't mean that you have to give that to them.

More than anything, Emotional Manipulators do what they do because they're lonely and insecure. The only way they know how to get attention is through provoking reactions in others, especially negative reactions. It may seem strange, but a powerful way to stop online manipulative abusive behavior sometimes entails doing absolutely nothing. Silence truly is an incredible communicator and sometimes the best "revenge" is giving no response whatsoever.

If Emotional Manipulators can't read you, they are left feeling frazzled, confused and powerless to control you. When they realize that their attempts to manipulate you have backfired, they will end up feeling anxious and abandoned. This is the first step at turning the tables around and regaining your power.

I'm a firm believer in surrounding yourself with only positive people while eliminating the negative people from your life; Facebook's "block" feature is especially valuable for this reason.

But if you insist on reading their posts, do so as a form of entertainment. Bust out the popcorn and enjoy yourself as you passively observe their many attempts to get your attention. See the posts for what they are: an expression of insecurity.

Once the behavior has stopped it is time to focus on moving on from the negative experience. When a Manipulator has turned your life upside down, it's completely understandable that you'll feel a lot of anger and resentment toward them, but in order to heal, you will need to eventually let go of your anger. There's a saying, "He who angers you controls you." This absolutely applies to Manipulators. The more you think about them the more they control you. Whenever you are ready, know that you can forgive and forget them without ever initiating conversation with them again.

Moving on is the act of releasing the desire to seek revenge or punish someone for an injury. Choosing to ignore toxic behavior does not mean you are a pushover. It means that you have instead chosen to refuse to act in an equally destructive manner.

You can't change your past, so any anger, resentment or regret you are holding onto is like, "Drinking poison, expecting the other person to die." It often takes a long time to get to a point when you can choose freedom over resentment. Sometimes your freedom can begin by uttering three powerful words: "I FORGIVE YOU."

chapter eight

Getting Your Facebook Fix

The first few days of my Facebook vacation were horrible. It was like being a drug addict. I felt lost. Everyone was on Facebook except me, and for the first time in a long time, I felt really anxious and alone. I kept thinking—what am I going to do now?

ADDICTION IN THE SOCIAL-MEDIA AGE

NOT SO LONG AGO, A FRIEND OF MINE TOLD ME THAT HER 11-year-old son ignores her because she spends the majority of her time staring at her mobile phone. He watches her scroll down her News Feed for hours and she gets annoyed when he interrupts her posts. She is always on Facebook. "I think I have a problem," my friend said. "I can't go more than twenty minutes without picking up my phone. I have to keep checking it." She spends half of her workday on Facebook and has tried several times to deactivate her account only to reactivate it within the same week. She checks it first thing in the morning and does a last update scroll before going to bed. She half-joking asked me if I knew of any Facebook Anonymous groups.

Currently there isn't an actual psychiatric diagnosis for Facebook Addiction—but there should be. Many psychologists are seeing more and more people like my friend who show the nine signs of addiction, which can occur either simultaneously or separately:

Preoccupation: You frequently have thoughts about Facebook experiences, whether past, future, or fantasy.

 Tolerance: As with any addiction tolerance, you feel like you need to spend more and more time on Facebook to get the same enjoyment or "rush."

 Chasing: You're overly focused on your posts soliciting responses or reactions from your Facebook friends.

 Risked romantic relationships: You spend too much time on Facebook or social media despite repeated requests from your partner, or you participate in questionable Facebook interactions despite risking or losing your relationship.

 Risked opportunities: You cannot get off Facebook long enough to focus on your work, school, or other opportunities even at the risk of losing them.

 Lying: You minimize or lie about the amount of time you actually spend on Facebook to your friends, family members, therapist, or coworkers.

 Loss of control: You've unsuccessfully tried to reduce the amount of time you spend on Facebook or find yourself unable to completely deactivate your account.

 Escape: You're spending time on Facebook and other social media as a way to improve your mood or escape problems, and you're finding that you prefer this escape to what used to be your escape.

 Withdrawal: At an extreme level, you experience irritability or restlessness during your attempts to cease or reduce Facebook usage.

These can range from mild to extreme problems. In my private practice I hear about Facebook-related problems at least once a day: someone is upset at someone else for what they've posted, someone is angry at their girlfriend for spending too much time focusing on her News Feed, or someone was caught on Facebook too many times and lost their job. When I've suggested to my clients who are experiencing Facebook-related problems that they take a break from it, they give me a look that says, *Are you crazy?* The mere thought of logging off, even for a weekend, often leaves them feeling nervous. This is the addictive element of Facebook—continuing doing what you love even when it is negatively affecting your life.

Loving Facebook isn't always a problem and can be a strength. It keeps us connected and informed. It gives us a forum to be heard. This chapter is not about giving up Facebook, or any other social media. It's to help you discover if you love Facebook too much, if you choose Facebook over other positive things in your life.

If you're not the one addicted to social media, I'm sure you know at least one person who's in need of a serious Facebook intervention. They comment on every post, share every single thought, post yet another profile pic and check Facebook 30 times a day or more. Whether they're asking you for another Candy Crush life or posting about their latest grocery store adventure, they live for what Facebook affords them: a way to be heard, an audience, a means for validation or attention, an instant connection with many people at the same time. So what's the hook of Facebook?

THE SLOT-MACHINE EFFECT

Facebook's News Feed functions like a Vegas slot machine: sometimes you win; sometimes you lose. But if you don't keep playing, when you leave the machine the next player might take your winnings. Similarly, sometimes there's an interesting Facebook update, sometimes there's not, but if you don't check Facebook you might miss one. Frequent Facebook visits cause something that psychologists refer to as *Intermittent Reinforcement.* Event invites, messages, and notifications reward you with a random "high," much like gambling. Just the anticipation of a Facebook response can be highly addicting.

Furthermore, we've become conditioned to post on Facebook to achieve a particular kind of feedback. Emotionally, we're rewarded with acknowledgement every time someone "likes" or comments on a post. If we don't receive any likes we hope we will next time. In real life, this compares to when we've received compliments about our work or wardrobe versus hearing jibes or jokes when we missed the mark. We like the feeling of excitement we get when we receive compliments or positive feedback. We do more of what got us compliments. When our updates are acknowledged we gain a sense of accomplishment, so it makes sense that over time we undergo a conditioning process where we become trained to check Facebook frequently. And that training causes us to lose power over our own decisions and behaviors.

Three factors—FOMO (Fear of Missing Out), checking-in, and sharing photos—appear to be the main culprits that get us hooked on Facebook, but the instant accessibility of the mobile device has further enabled our addiction. We have at our fingertips the ability to freeze a moment in time, share it, and then monitor how many people are responding to our moment. The mobile device has even made it possible for us to meet all of our sensory needs at once: we can grab our mobile devices and hear someone's voice as we watch their facial expressions on FaceTime; we can text someone, take a

Douglas, 52

Geneseo, Illinois

I like to share photos of my vacations or business trips. Quite often I'd post a photo of a place I've been to with a caption describing my impressions of the beautiful scenery, the crowds, and the food. Once I posted a photo I'd feel a little nervous wondering if anyone would comment on it. After five minutes, if no one reacted to it, I'd wonder why. So I'd post another pic—one that was brighter or happier. I'd get excited once I received an email notification letting me know that someone posted a comment. It didn't matter if I was in a meeting—I'd check it. I think I need to know what people post about me. I want to make sure it's positive. How long could I go without checking a Facebook response once alerted to it? I don't know... maybe a few seconds.

☺ Like   💬 Comment   ➞ Share

pic of that same texting conversation (as it's happening live), then plan a lunch date online and find a restaurant on map search. And once we decide what we're going to do, we post it on Facebook.

FOMO

"FOMO," or fear of missing out, is widespread in the social-media culture. If you're not keeping up-to-date with everything that everyone is doing, you feel left out. FOMO is certainly not a new concept, but it's hard to resist when you have instant access to what everyone else is doing. Facebook, Twitter and Instagram have all made it possible for us to instantly see everyone's fabulous life: where they're hanging out, who they're hanging with, and what they're eating (Instagram people, we get it; you like food). Is FOMO having a negative impact? Fear of anything is not especially healthy for anyone, and it's definitely one of the concepts behind Facebook addiction. Fear often leaves us feeling trapped or blocked from pursuing things that we really want to do. When we miss a friend's posts, we miss out on information. When we miss out on information, we feel regret.

BEYOND A 'LIKE' OR A 'FOLLOW'

With the explosion of smartphones most Facebook users have turned to the Facebook mobile app. One super trademark of the Facebook mobile app is the ability to "check-in" due to built-in GPS system.

Through Facebook, we (along with many corporations) can now keep track of a person's movements. This is one of the major disadvantages to the location feature—namely, when you check into a location, you instantly allow others access to where you and/or your family members are that very second. Aside from the obvious safety concerns that many of us choose to ignore, we're also willingly allowing third parties, individuals and corporations to know a whole lot more about us. What's fascinating, even more than our need to watch other's movements, is the emotional need to be watched and tracked.

Social media has influenced us to assume that our friends need to know where we are—and indeed, many of them do want to know. When exactly did tracking our friends become a social trend? Did Garmin or Google, innovators of GPS technology and mapping, anticipate this? Perhaps we all have an instinctual urge to know where people are.

People love the checking-in feature because it adds another element to their personality. As with most announcements on Facebook, the places where you check-in let other people know what type of person you are. The places you hang out can be a reflection of your personality as much as, say, what you wear or what you say. Are you more likely to check-in at a place that serves an amazing brunch or the best mojitos in the city? Are you a happy hour kind of gal or one that can't wait to get home after work? Checking-in is different than your regular Facebook status because there is a level of accountability attached to it. In order to check-in somewhere, for the most part, you have to actually be there physically, so it forces you to be more truthful than just a regular status update.

Maybe you want to locate friends, maybe you want your friends to locate you, or maybe you want to mark your territory, so to speak. Whatever your reasons, checking-in is a very popular feature that gives you a psychological sense of achievement when you can announce it to others. Facebook counts on this feeling: The more reinforced you become with checking-in through likes and comments, the more you'll do it. The more you check-in to places, the more third-party corporations will love you. What do we do about this vicious cycle?

Everyone's a Photographer

Nothing says more about you on Facebook than the photos you choose as your profile picture and cover. Sharing photos expresses who you are, especially when you share those of yourself in action. Action photos are far more interesting than your posed portraits because they represent your life as it is happening at that very moment. You're at a bar hugging your friend, you're running with your teen at your neighborhood 5K run, you're holding up an award after receiving it, or taking a photo of yourself in your car as you're driving. This is you in motion; this is how you behave in real life, or at least how you wish to appear. We not only want to show people what we do, but also how we do it. Our status updates are just words but photographs are more believable, and many people count on this.

Sharing photographs might not be such an addictive activity on Facebook if it were not for smartphones. It's just so easy. We're now armed with the ultimate handheld super device: telephone, recorder, flashlight, GPS, camera—you name it. The instant accessibility to your friend's updates and to your own wall encourages constant updates.

Facehooked

A clear sign of addiction is when a person wants too much of something and they just can't stop, regardless of their conscious knowledge of such things. Are we receiving too much information or too little in today's technology culture? Is the speed of information and how we process it going to continue at its current pace? What if it's never enough? How much information can our senses take? We've gotten so busy focusing on whether or not we can gain more and more online information that we've forgotten to question whether or not we *should*.

Sorting through the (almost) endless iPhone system and application upgrades gives us plenty to do and yet we desire more. The more options we have, the more we want, and the more we want, the more we become addicted. Facebook has tapped into a very powerful and seductive combination: instant access, sharing photos, and tracking others. The fear of missing out on these very things leaves us feeling hooked. If we can manage these three things separately, we might be able to use Facebook in a fun, healthy and productive way; however, in combination, all the elements may be too much for some people to handle.

There is such a thing as technology overstimulation. Some people simply can't keep up with the pace of today's world. As technology continues to advance, our ability to process everything that we're being hit with will become affected. The social-media sensory overload we're all receiving may be leading to increased levels of anxiety and mental instability in today's cultural climate. The speed of our culture is affecting the speed of our lives in terms of our ability to connect, touch, feel, and process what is happening right in front of our eyes.

As with most addictions, insight is the first step to recovery. Before you can treat a problem you have to first realize that you do, in fact, have a problem. Are your personal relationships taking a backseat to Facebook? Have you ever concealed Facebook use? Do you stay on it much longer than you intended? Do you think about Facebook even when you're offline? If you answered yes to any of these questions, you might be a Facebook addict.

Entitlement and Addiction

I used to work at an art college as Director of Counseling. One of my roles was to meet with students and faculty in order to help them navigate and resolve certain conflicts. One day a student was sent to the Dean's office because she refused to get off Facebook during class. The professor had instructed all her students to leave their current classroom and head over to the drawing room. Despite repeated requests from the professor, the student remained fixated on her laptop and refused to stop typing. This was the student's response when I asked her why she believed she was sent to my office:

> I don't know why I'm even here. I was in the middle of something on Facebook and [the professor] wouldn't leave me alone. She should be charged with harassment. I don't see what the big deal is. She could've waited a few more minutes. I mean they get paid whether or not we show up to class or not right? So she should just focus on her job and mind her own business.

The student ran afoul of other instructors due to similar behavior and was eventually expelled from the college. Clearly, this student had more behavioral issues than a serious Facebook addiction, but I was struck by her perception that her Facebook posts were just as important, if not more important, than her academics.

Aside from my work at this college I also taught Doctorate-level courses at a private graduate psychology school. You would think that at the graduate level, students would not be so prone to display entitlement or smartphone or Facebook addictive behavior, but in certain cases, I had to reprimand students for not being able to put their smartphones away or when I caught them checking Facebook during lectures. I have no doubt that some of the students posted nasty comments about me on Facebook, and I had no problem with this—as long as it was after class.

I have learned that the best way to avoid Facebook and smartphone addiction in classrooms is to let the students know at the onset of class that they will be penalized a grade point every time I see them using their smartphones. In my private practice I sometimes have to ask my clients to please put their smartphones away during session. I have spoken to other therapists, medical doctors, and even schoolteachers, who are trying to have a conversation with parents. They say that at times they don't have the full attention from their audience due to smartphone and push-notification distractions.

The power of recent technological advancement is awe-inspiring, but we haven't thought through the affects such connection will have on individuals and communities.

A CLEVER DISTRACTION

Social media has become the newest activity with which to distract ourselves. Whether it's video games, PC gaming, Facebook, or Candy Crush, many people spend vast amounts of time participating in these distractions—far more time than intended. Many of us get so absorbed we program our iPhones to push messages instantly, notifying us of an update, message, or photo tag. In fact, many people were upset that push-notifications on iPhones were not instant as on Blackberry devices. Apple had to figure out a way to instantly push updates so they wouldn't lose business.

What would it mean if technology didn't push instant notifications on us? Would we be just as addicted to checking Facebook, instant messages, and tagged photos? The technology behind the mobile device may actually trigger the addiction. Even if we don't feel a need to check-in, mobile technology and Facebook are pushing the addiction on us. Most apps will push updates unless we actively change our settings so they don't. Our mobile devices are the ultimate enablers of addiction. An alert on our phones can cause many of us to check out of our real life in order to check-in on our Facebook life. The distraction is overtaking the reality. The most compelling question then is why are we so drawn to escaping reality?

Roberto, 43

Chicago, Illinois

I spend a lot of time on Facebook. I check it, on average, every 15 minutes. I spend my day searching online for funny posts to share with my friends. I need to one-up myself. Most of my stuff is really funny, so I keep looking for better things all the time. I believe I'm just as funny in real life, yet the reactions I get on Facebook matter more to me than the reactions I get in real life. I know I'm funny. Everyone knows I'm funny, but now every time I post, it has to be better than the last one. I have a Facebook reputation to think about. Ha! People seeing that you're funny online just feels better than people knowing you in person. It's like a status thing. My friends look forward to what I post and I have to deliver.

     👍 Like   💬 Comment   ➦ Share

Are we driven to develop our social-media selves because of a deep-seated need within us, or are other Facebook users enabling this new behavior? All Facebook users can be prone to compulsive posting. We're experiencing a collective need to share our lives, and the instant gratification we receive when we see our lives (or the lives of others) played out is intoxicating.

Facebook users are now basing every decision—where they go, whom they associate with, what to wear—on how they appear to others online. We're feeling more and more pressure to perform and using Facebook as a vehicle. We want everyone to see us, and this want, this strong desire, is the driving force behind every post.

What's the Brain Got to Do with It?

So is Facebook controlling us? It's complicated. Many research studies have shown that people who spend a lot of time on Facebook tend to experience more depression, more anxiety, and lowered self-esteem. So why aren't we making efforts to pull it back a bit? Because there is, without a doubt, a seriously addictive component to Facebook. A study from Retrevo, a consumer electronics site, asked social-media users how often they checked sites like Facebook and Twitter. They found that our brains release a burst of dopamine whenever a post captivates us. This causes a feeling of euphoria. We get hooked on that dopamine high and keep visiting Facebook to chase it.

Like a drug controlling the drug addict, we love Facebook and we hate it. We want to spend less time on it and yet we can't. We can't wait to post every marvelous experience we have. We need our Facebook fix. And until we can find some other way to keep our minds occupied, Facebook will continue to be our drug of choice. Not only is Facebook changing our choices, but it's also changing our brains.

Medscape.com reported findings from a recent literature review on Internet Addiction Disorder (IAD) presented at the American Psychiatric Association's 2014 Annual Meeting. This review showed that Internet addiction is linked with certain brain abnormalities and changes in blood flow. Although IAD is not currently an established disorder, research is showing that it is associated with certain symptoms such as marked distress, mood changes, tolerance, withdrawal, and impairments of social, occupational, and academic performance. It also shows that IAD is associated with depression, suicidal behavior, obsessive-compulsive disorders, eating disorders, attention deficit/hyperactivity disorders, as well as alcohol and illicit drug use.

Sree Jadapalle, MD told reporters, "Increased blood flow is actually seen in the areas of the brain involving reward and pleasure centers, and decreased blood flow is observed in areas involved in hearing and visual processing." This means that the more time you spend logged on, the more your brain will focus on the pleasure areas of your brain, and less on the areas of your brain that keep you safe and alert, such as hearing and vision. Dr. Jadapalle added that the prevalence of IAD among American teens is about 26.3%, which is more than alcohol and illicit drug-use disorders.

The research also shows that Internet addiction is leading to heightened sensitivity to the reward areas of the brain and decreased sensitivity to monetary loss. Therefore, if you're prone to spending a lot of time on Facebook, you will likely not care about the consequences of your addiction, including emotional, social, and even work-related problems.

THE PSYCHOLOGY OF ADDICTION

Nearly all addictive behaviors are triggered by emotionally meaningful events. This means that something in our past has left us feeling helpless in certain situations and we turn to other distractions for relief. Many of us feel a strong pull to do something to feel less out of control and for Facebook addicts this entails checking and rechecking their News Feeds. If Facebook addicts suddenly stopped checking Facebook, they would likely choose some other addiction—gambling, video games, drugs or alcohol—for someone who is avoiding dealing with a painful event, engaging in an addictive behavior feels like a better solution.

A feeling of anxiety or helplessness precedes almost all addictive acts. It goes like this: I feel anxious; I need something to alleviate my anxiety; I will do something that reduces my feeling of anxiety. A smoker may feel that a cigarette instantly gets them out of an emotional jam. Facebook is usually not as hazardous to one's health as something like excessive smoking or drinking, but the end result can be the same.

Psychologically, addiction happens when we're so focused on something that it takes over our world. The drive to feel better is so strong that we turn to addictive behavior hoping to feel less vulnerable. Initiating the addictive act (or even thinking about it) creates a false sense of control over negative emotions. Plainly stated, addictive acts seem to make whatever is bothering us at the time, better. Facebook can alter how we feel about ourselves, and when we perceive our lives as problematic or stressful, Facebook provides an excellent escape. However, when we escape into the Facebook world for too long, we can lose sight of other important aspects in our life: family, relationships, and real-life friendships.

Why would we do something that isn't good for us? What is it inside of us that pushes for unhealthy vices?

What most addicts don't realize is that the very things they turn to for "help" are the things causing the most stress. Maybe smoking helps you through a rough patch. No matter how bad you feel, a drag will help. Until it doesn't. Then maybe you find yourself desperately scrounging cigarettes off strangers, or searching for a "clean" butt somewhere.

When we repeat addictive behaviors, we get lost in the behavior itself. It leads to a loss of good judgment and, at extreme levels, an increase in destructive behavior. People become addicted to gambling because initially they get feelings of joy associated with winning a game; it also serves as a distraction from life. But if gambling is the way a person regularly deals with stress, then he will play—over and over again—with an intense and irrational drive. A behavior becomes a compulsion.

Compulsions aren't voluntary and they definitely don't give pleasure. Instead, a person with a compulsion engages in a particular behavior to relieve discomfort, which becomes overwhelming if an activity isn't repeatedly performed. The compulsive act relieves distress but only temporarily, and compulsions do not relieve or prevent the fears that inspire them. For example, you might feel obliged to check that your stove is turned off every few minutes. Even if you are certain that you turned off the stove, you may already be conditioned, or trained, to do so.

THE SKINNER BOX

B.F. Skinner, a psychologist and behaviorist, believed that any human action is performed to achieve an expected result. For example, if I perform an action and the consequences of my action are negative, there's a high probability that I won't repeat it; however, if the consequences of my action are positive, then my actions will be reinforced. He called this the *Principle of Reinforcement.* In order to prove this theory, Skinner invented what's known as The Skinner Box. In this box, or chamber, he studied conditioning by teaching an animal to perform certain actions (like pressing a lever) when presented with a light. When the animal correctly pressed the lever, food was automatically delivered as a reward. The mechanism delivered a punishment (no food) for incorrect or missing responses. Eventually the animal was conditioned to press the lever again and again when it was stimulated with the light.

We can be conditioned too. I once owned a rabbit. Every night at 2:00 a.m., he would thrash around in his cage waking me up. To get him to stop making noise, I would let him out of his cage. Once he gained his freedom, he enjoyed roaming around my kitchen all night. Because I had let him out of his cage a few times (reward), he learned that if he thrashed around and made noise, the reward would be repeated. The rabbit had trained me to let him out of his cage.

We all engage in certain behaviors for rewards or to avoid negative consequences. Compulsions can act as rewards: They distract us from negative feelings and instead of dealing with our feelings directly, we engage in *displacement;* we place all of our angst, fears and frustrations into something else—an addiction. In Freudian psychology, displacement occurs when our minds unconsciously substitute a new object, or distraction, for other objects that we perceive as uncomfortable or dangerous. Facebook has become the ultimate compulsion and displacement for emotions. People addicted to Facebook compulsively check the site as a distraction and their feelings can be displaced where they are expected to appear: on a Facebook wall.

WHAT TO DO IF *YOU'RE* FACEHOOKED

Instead of going on Facebook and doing what you usually do, start asking yourself what you're really getting out of it. How valuable is it to your own life? What's missing from your life? What's making you spend so much time on Facebook and away from your real life? Do you have too much free time? Do you live somewhere you just don't know anyone? Are you just escaping? A food addiction

Beth, 36

Grand Rapids, Michigan

I was staring down at my smartphone so much that I caused myself neck pain. I'd forgotten what it felt like to look up and around. People had to tell me to put my phone away and every time they did I'd get angry. I love my phone and scrolling through my News Feed, but Facebook can't be all that great when the third guy in a row has dumped me because I couldn't put my phone away. My therapist said that I was using Facebook to sabotage relationships because I was hurt in the past. Some of this might be true. Friends on Facebook don't disappoint you as much as guys can.

     ⸺ Like   💬 Comment   ↪ Share

has very little to do with food and often reveals itself because something is wrong in your life. Similarly, a Facebook addiction has very little to do with Facebook and is a sign that you might be avoiding something in the here and now.

If this sounds familiar, you may need to make some serious changes in your life.

FACEBOOK IS NOT THE PROBLEM; FIND OUT WHAT IS

Many people are able to enjoy many of Facebook's benefits without feeling overwhelmed and can function with it in their lives just fine. The main problem with overuse is when you cross the line from casual social networker to dysfunctional behavior. If you have a Facebook addiction, you tend to spend way too much time on it because you feel like you're missing something. Begin by asking yourself how you are feeling on most days. Are you depressed, nervous, mourning the loss of something from your past, or not at ease in your life? If so, here lies your main emotional trigger. It's easy to blame the obvious point of our attention (Facebook), but often an addictive behavior is just a symptom of some other underlying problem. Take some time to evaluate if you are using Facebook as a bandage to your life. Are you using it to avoid dealing with certain issues in your relationship or job? Focus your attention on fixing whatever is wrong or missing from your life. Once you are aware of the underlying issue, you can be more confident in managing your Facebook use.

When I ask people how often they check their Facebook News Feed, many say, *Oh, a couple of times a day.* Upon examination, it often turns out to be far more than that. Insight is the first step to changing behavior; keeping track of how often you actually check your News Feed may help you "see" your pattern. Maybe you log on only in the mornings; maybe you're a binge user, checking it only once a day but staying on it for hours on end; or perhaps you need to compulsively check it a dozen times a day. And even when you don't log on, your phone or tablet may send updates triggering you to log on. Once you're aware of your behavioral pattern, you'll be able to identify a trigger that leads you to log on in the first place. This could be as simple as boredom at work or something far more serious. Awareness of what you're doing, and at what time you're doing it, is useful in determining purposeful Facebook use versus using it mindlessly. Write down these triggers so that you can become more aware of them—in order to avoid reacting to them.

TAKING A FACEBOOK BREATHER

A couple of years ago I attended a retreat in Arizona. The first thing I had to do when I arrived was hand over my cellphone. You have no idea how anxiety producing this was for me. I reluctantly complied, but after a while I realized something: living without social media for ten days is pretty amazing. I found myself spending more time communicating with the person seated next to me. I listened more intently, and spoke with more meaning.

Try leaving your phone in your bag for an hour or leave your tablet at home for a day. As a reward for taking a small break from technology, spend that time cooking something wonderful or doing something social with a friend or family member. Designate some time, whether it's a few days or more, away from Facebook and other social media. Let your friends know that you'll be away "working on a project" and then just log off. Taking a break from anything gives us a new perspective. Finally, if you're not ready to quit, you can always join an anonymous group for addiction on Facebook itself. Believe it or not, Facebook has over two hundred anonymous addiction groups, though this might not work so well when your goal is to take a break from Facebook.

chapter nine

You Like Me . . . You Really Like Me!

Getting likes on Facebook was fun at first but truthfully, now they're necessary. If I post something and I don't get enough likes, I freak out—kind of like I just made the worst mistake in the world. If no one likes my pic I either need to quickly delete it or try to get people to like it. Otherwise it shows my mistakes; it makes them obvious.

SEEKING APPROVAL AND VALIDATION ON FACEBOOK

WE WANT PEOPLE TO LIKE US. BUT NO ONE SHOULD EVER be able to dictate your value. That's like going up to a total stranger and asking, "Hi. Can you please tell me how much I'm worth?"

Hearing input from others seems like a good idea. There are times when we think that we simply can't be as objective as we want to be so we'll rely on others to shape our values. The problem is that we can't always tell which opinion is objective. When we rely on others' opinions too often, we can be led to place more value on their opinions than our own. Where does the drive to seek outside validation come from?

As infants, we depend on others for nurturing, protection and support. As children, we learn to seek approval and permission from our parents or guardians. We care what they think so it becomes important for us to get positive feedback from them. We feel safe knowing that we are behaving correctly, and who better to ask about than the people who are supposed to look out for us? Over time, we have learned that we should seek approval from others as well. For example, if our guardians were not around to correct our behavior, then our teachers took on this role, or our supervisors. Throughout our lives we continue to ask people if we are making the right choices. We have a strong emotional drive to have other people approve of our choices, and aside from agreeing with us, perhaps what we really need is to feel that people understand us. How does this relate to Facebook and social media? The "like" feature.

The ability to like someone's post is a powerful Facebook feature because a lot of Facebook users unknowingly depend on it. We do not post things just to post them. What would be the point? The allure of the like feature is much more than our friends endorsing what we post. Receiving likes on Facebook sends a deeper message—it conveys that our friends endorse and celebrate us.

Brenda, 25

Phoenix, Arizona

Of course I 'like' my own posts! I'm the one that posted them. I also 'like' them so they'll show up higher on the News Feed. Most of my posts are funny sayings or flattering selfies that show off my abs. From time to time, I'll share an update about myself, or my kids, but mostly I shares things that I think my friends will 'like.' It annoys me when people don't take the time to press 'like'. I mean, that's what you're supposed to do on Facebook. Not for nothing, but I worked hard to look like this, and if I take the time to like your post, the least you can do is return the favor. That's what good friends do. When I don't receive enough likes for my posts, I ask my best friend to like them for me.

    ◦ Like   💬 Comment   ↪ Share

There is nothing wrong with sharing accomplishments with those close to us. We should strive to improve ourselves. One way to receive perspective on our improvement is through feedback. This makes sense. Maybe it's not so much about asking for feedback as it is about boosting our sense of self-worth. Emotionally, feeling approved of makes us feel more confident. We all seek approval some of the time.

Gaining acceptance, more than any other factor, directly influences how we measure our own value. This is why we put so much value on other people's opinions. The irony, of course, is that this dependence upon external approval is actually the killer of self-esteem. When we rely too much on approval, we give away our power; we end up trying to improve ourselves for other people, which in turn causes us to lose all sense of self.

The Ego Boost

The validation that we receive on Facebook is addicting. Think about it: When you post something on your wall, you're likely to receive a response containing some form of praise. Close friends and family members are supposed to give us a little "push" when we're down and try to help us feel better about ourselves. When people post new profile pics, almost immediately their friends will click the like button or express a compliment. This is not much different than someone telling you in person that your shoes look great or that you look incredible. Validation on Facebook seems harmless enough and can be beneficial, especially if we don't find validation elsewhere. A few kind and supportive words can provide us with positive feelings and give us a glimpse into new possibilities about ourselves. It can also motivate us to improve our current circumstances.

Ideally, there are plenty of people we can turn to for support and validation. Aside from our guardians, we have our other family members, extended family members, and friends. In a dream scenario, they would be available to provide us with reassurance when needed. We tend to respect the opinions of those to whom we feel connected and trust that they have our best interests in mind. We also hope they can provide constructive and helpful feedback.

Lola, 30

New Haven, Connecticut

I had been in a twelve-year relationship when I first joined Facebook. I had always been very popular in high school, but after I got married my husband and I were homebodies. I was always home; I had zero life. I had no social activities outside of my marriage. I think that I kept myself sequestered because I was very self-conscious about my weight after I had my child. When I got on Facebook in 2009, I suddenly had many friends. There was always someone to let me know that I looked beautiful. Some men even added comments letting me know how sexy I was. My marriage was failing, and Facebook gave me a reawakening and an understanding that maybe there's a whole life out there that I was missing out on, especially when I started to receive invitations to many parties and gatherings. Facebook helped me feel better about myself.

🖒 Like   💬 Comment   ➦ Share

Validation is a needed and beneficial thing. Sometimes, however, we lack people in our life who provide us with support and when this happens, Facebook can be a worthwhile alternative.

Healthy Facebook Validation

When people go through challenging times, one way they can cope is to share their experience with those around them. When life becomes especially difficult, they may turn to a loved one to hear them out; there is something emotionally soothing about feeling understood. This is how a lot of people cope with stress. In fact, this is one of the foundations of healing within psychotherapy.

People go to therapy for many reasons, but the one consistent element of most forms of therapy is that someone will be there to listen to you talk about your world and make efforts to understand your place in it. No matter what kind of treatment techniques are used, many clients report that what really helped them is feeling connected to their therapist. This sense of connectedness happens when clients feel that their therapist can empathize and relate to them on their level. Most clients need to know that their therapist "gets" them before the real therapeutic work begins.

Many of us need to share our problems with someone because we don't like to go through hard times alone. The problem is we can't always get our messages across in regard to our validation needs. We perceive certain situations with clarity. When others do not see them in the same manner it can be difficult for us to feel heard. This confusion may lead to anxiety, so we try harder to get them to understand. When this doesn't work, we may rely on the power of the masses; we will share our problems with several different people in the attempt to increase the chances of being validated.

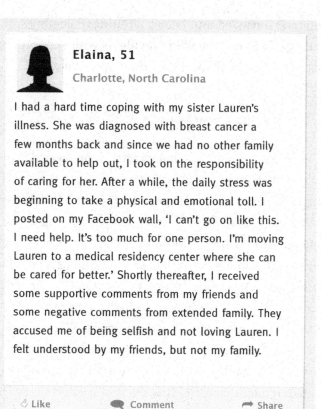

Elaina, 51

Charlotte, North Carolina

I had a hard time coping with my sister Lauren's illness. She was diagnosed with breast cancer a few months back and since we had no other family available to help out, I took on the responsibility of caring for her. After a while, the daily stress was beginning to take a physical and emotional toll. I posted on my Facebook wall, 'I can't go on like this. I need help. It's too much for one person. I'm moving Lauren to a medical residency center where she can be cared for better.' Shortly thereafter, I received some supportive comments from my friends and some negative comments from extended family. They accused me of being selfish and not loving Lauren. I felt understood by my friends, but not my family.

    👍 Like   💬 Comment   ➡ Share

In her heart, Elaina knew that that she was doing the right thing. She was not abandoning Lauren. She planned to visit her every day, bring her whatever she needed and planned to support her in every way possible. The comments from her family, however, caused her to question her own motives. With limited support, Elaina turned to Facebook. Seeing the abundant number of "likes" and supportive comments helped lift her spirits. She especially enjoyed reading her friends' posts that defended Elaina against her relatives' judgmental statements. This is an example of how Facebook can be a safe place to turn to in terms of us getting the reassurance we sometimes need. Facebook also provides us with a platform in which to be heard and help others.

Is Facebook Validation Really Validation?

The simple answer is yes. Facebook provides many of us with a sense of connectedness and understanding. If we didn't feel understood or validated through Facebook, we'd never even start posting updates. Facebook increases our visibility and feeds our desire to be heard. With minimal navigation and a few clicks, we can hint at our uneasiness on our Facebook walls and receive immediate affirmation from supportive friends. However, our ability to feel understood through our online interactions is limited, because the part of us that gets understood is the part of us that we choose to share on our walls.

Let me be clear: I'm not saying that receiving texts and social-media commentary don't help us to feel connected to others; they do, but only to an extent. We are complicated and sensitive creatures living in an increasingly complicated world. On a daily basis, through our five senses we take in information from our environment, learn by example, and hopefully gain some wisdom and grow as human beings. Our sense of being understood does not develop through online interactions alone, because staring at a screen does not provide us with the full range of emotional support that we need. We need the stimulation of a loving parent, the sensations of a friend's touch, or the memory of a lover through smell. Interacting with someone online can never produce such experiences.

Many argue that they find Facebook validation more important than real-life validation. Have you ever called a friend to wish her a happy birthday, and had her ask why you didn't just post your greeting on her Facebook wall? Many of us now prefer public endorsement over personal interaction. As a society that is quickly becoming entrenched in social media, we are experiencing a growing dependency on such public validation. Are we slowly becoming a collective culture of validation seekers? If so, how does this affect us?

A dependency on online approval makes us lose our sense of self-reliance. As we continue to partake in the giving and receiving of Facebook validation, our behaviors will change. We will stop relying on our knowledge and intuition to get us on the right path and instead look to others for direction. We will stop noticing our strengths and instead feed off of our insecurities. We will stop living for the moment and instead live in the Facebook moment.

UNHEALTHY FACEBOOK VALIDATION

Sometimes people are inclined to argue against the compliments they receive, feeling like they accomplished goals because they "got lucky" (not because of their hard work or intelligence), undermining their strengths. They engage in a lot of self-blame. They find it difficult to take ownership of the positive aspects of their life and have a hard time looking at themselves objectively. They also tend to place more value on other people's so-called objectivity because they do not trust their own judgment. Even when they feel happy or proud of themselves, they cannot enjoy such feelings until someone else expresses positive sentiments toward them.

Hannah, 37

Columbia, South Carolina

Whenever Judy likes or posts a positive comment on one of my photos, I feel less anxious. I spend a lot of time browsing her Facebook wall. Judy is my boyfriend's ex. She and I get along, and Judy is married with four kids and appears happy in her marriage. I keep posting photos of me and my boyfriend on Facebook, hoping that Judy will like or comment on them. I know it doesn't make sense, but I need her to see that we're happy. I really want her to comment on the pictures too to show me she gets it. It drives me crazy sometimes, but I can't help it. I'll think about it all day if she doesn't like our photos.

    Like   Comment   Share

In cases like Hannah's, validation-seeking behavior has very little to do with the reactions of others, and has everything to do with insecurity or low self-esteem. The focus is external, yet that's not where the problem lies. Giving tons of Facebook likes will not increase your sense of self-worth, and receiving tons of likes will not make you feel more connected. Perceiving Facebook validation as a necessity means that you've become dependent on it.

Somewhere along our life journey we learn a lot of lessons. We can accept some lessons more easily than others. One of the greatest lessons we can learn is that we can't control the thoughts and behaviors of other people. We can only learn to control our own thoughts and reactions. People will think what they want to think, and, try as we might, we cannot please everyone—nor should we try. Placing too much emphasis on what other people think, seeking validation from them, and depending on outside opinions to determine our self-worth is exhausting and pointless.

You must stop trying to please everyone because you can't possibly be all things to all people. You've heard this before but what does it really mean to you and your interactions on Facebook?

Basing your decisions of what to post on Facebook for the approval of others gives them power over your happiness.

Constantly striving for approval means you do not approve of yourself. And if this is the case, it doesn't matter if you receive the approval of ten thousand Facebook friends because emotionally it will never be enough. Rather than trying to get your needs met by others, make it clear that you approve of yourself.

Ironically, the greater your need for attention and validation, the less people will want to be around you. People tend to gravitate toward those who display self-confidence. People with self-confidence do not require or request approval. They also do not need to show off or mock others in order to feel good about themselves. Try posting because it genuinely makes you smile, reply to your friends' posts in meaningful ways, and don't take the comments to your posts too seriously.

Believing that you need other people's validation to feel good about yourself only encourages your own feelings of worthlessness. Covering up your feelings by focusing on the reactions of others will not help overcome the feelings you are trying to hide. The only way to get rid of troublesome feelings such as hopelessness or worthlessness is by understanding why you feel this way. You may be able to discover this on your own, or you may want to see a therapist for help. Knowing why you are experiencing self-doubt or low self-esteem is a first step to ultimately conquering these negative feelings.

Five signs that you may have become validation dependent:

You spend a lot of time editing your selfies on Facebook in the hope that they look good enough to trigger likes or positive comments.

You spend a lot of time thinking that you should be stronger, braver, prettier, thinner, or smarter and make efforts to embellish certain details about your life on your Facebook profile.

The more Facebook friends you have, the better you feel about yourself.

You seek support from your friends and Facebook friends, and the more you want support, the less you seem to get.

You have a hard time hearing constructive criticism, and when you receive a negative comment on your Facebook post, you quickly delete it.

Where Do We Go From Here?

During those difficult times when you are in need of support, first figure out exactly what you need. Asking for support and encouragement when we need it is healthy and will help others understand how they can help. You might even want to express exactly what you're looking for and what you're not: *I really need you to listen to what happened today at work. I don't need advice just yet—I just need you to listen.*

Identify the people who matter the most to you in your life and who have, in the past, provided you with kind and objective feedback. Think about the people who motivate you versus the people who exhaust you. These supportive people are the ones you should turn to for advice and constructive criticism. Looking for support from strangers or from those who haven't provided you with sound advice in the past doesn't make sense.

Finally, remember that gaining praise or attention on Facebook has nothing to do with your personal value. You don't need to earn your value as a person through the number of likes that your posts receive. Receiving affirming or positive statements can improve your mood, but you can't rely on this kind of attention to fulfill your emotional needs. Instead, practice letting go of seeking validation for your choices and instead learn to accept yourself as you are. When you make a decision, check to see if it feels right for you. Does your decision make you feel happy or more present? If so, you're likely on the right track toward achieving an even greater sense of authenticity and confidence. How can you be more authentic?

When you stop looking for validation, you may find that you already have it. When nobody validates you, learn to validate yourself. When nobody compliments you, compliment yourself.

Encouragement and motivation is found within, and one way to change how we encourage ourselves is through the words we use.

Notice how you express yourself on Facebook. If you find that you put yourself down, make efforts to catch yourself doing this and stop. Likewise, when you find yourself posting in a way that makes you feel proud and good about yourself, acknowledge it and try to do this more often. Giving yourself credit for your positive qualities is not egotistical. Honor yourself for all that you have overcome and accomplished and know in your heart that the only person who needs to accept you—is you.

Approval and support from those we trust can, at times, be a mechanism that carries us forward to a place where we can discover our authentic and sincere lovable self, but validation from those who are not looking out for our best interests does not help us in any genuine way. Once you realize the difference, you become free to live the life that you want to live and you will do so for yourself—not for other people.

chapter ten

Checking In Without Checking Out

I won't accept someone I'm dating as a Facebook friend until 30 days. I've learned my lesson in the past. I want to get to know you—not the person you want me to think you are. The only way to do that is to stay off of Facebook. Someone told me once that you shouldn't really text either. Not for a while anyway. This makes sense and so far, it's worked for me.

Using Facebook Without It Using You

THROUGHOUT FACEHOOKED, WE'VE FOCUSED ON HOW Facebook affects us on an individual level. But Facebook is used across the globe, and other countries are using Facebook in unique and inspiring ways. At its best, Facebook is a powerful tool for change, which is perhaps best seen in other countries' use of Facebook. There are now more than one billion Facebook users across the globe. In countries such as China, Syria, and Pakistan, access to Facebook is restricted thanks to social media's reputation as a breeding ground for political rebellion. So if users are going to use Facebook, they must be creative, using proxy servers to access Facebook in order to simply gain access or learn what the rest of the world thinks about events happening inside or outside that user's home country.

In 2010, an Egyptian businessman in Alexandria, Mr. Khaled Said, was pulled from an Internet café and then beaten to death by police. Activists believed he was killed because he had posted a viral video with evidence of police corruption. Within days of his death, an anonymous human rights activist created a Facebook page against Egyptian police—We Are All Khaled Said—where he posted cellphone photos of Said's battered face and YouTube videos that displayed contrasting photos of him before his death with bloody images from the morgue—130,000 people liked the page that month. It quickly became the largest rebellion Facebook page in Egypt (and continues to spread the word about demonstrations in Egypt), which ignited a massive uprising. Social networking offered Egyptians a forum in which to bond over their outrage, and to organize and mobilize against government abuses.

Facebook is an easy target for governments looking to control the masses, and these incidents are not always occurring in the countries you would expect. In 2009, Massimo Tartaglia attacked Italian Prime Minister Silvio Berlusconi by hitting him in the face with a toy model of Milan's Duomo Cathedral. Berlusconi received little sympathy. Facebook users went online to praise Tartaglia, one group saying: "We condemn all forms of violence, but the sight of Berlusconi's battered face is priceless." There were online pledges in support of Tartaglia and calls for him to be elevated to sainthood. Furious, the Prime Minister's ministers threatened to shut down anti-Berlusconi Facebook groups. It would not be the last time Italian Facebook users sent a political message to their leader. A Facebook group later emerged called "Let's Kill Berlusconi."

These incidents illustrated that the Italian government not only did not have a firm grasp on the importance of social media in the minds of the Italian people, but they also helped individual Italians see how their fellow citizens truly viewed the Prime Minister. In this way, these events were both innately personal experiences for those involved, and at the same time, national and global events. They were also a chilling reflection on how Facebook users do not always feel the need to filter their emotions or thoughts online.

The Top 10 countries with the most Facebook users, according to TechElipse.com, as of July 2014:

1. **United States of America:** The total number of Facebook users in the U.S. is approximately 159 million. Top users range in age from 18-34.

2. **Brazil:** Total number of users is approximately 72 million.

3. **India:** At No. 3, India ranks on this list at 67.5 million users.

4. **Indonesia:** Ranking fourth, Indonesia's users total 48.8 million.

5. **Mexico:** At No. 5, Mexico! With approximately 42.5 million Facebook users, social-media presence is seen here as rapidly growing.

6. **Turkey:** Approximately forty-four percent of the total population in Turkey (32.7 million) use Facebook.

7. **United Kingdom:** The UK, comprised of England, Wales, Scotland and Northern Ireland, together boast approximately 31.1 million Facebook users.

8. **Philippines:** Approximately 30.2 million people use Facebook in the Philippines.

9. **France:** There are 25.3 million active users in France.

10. **Germany:** One out of every three German citizens uses Facebook, totaling approximately 24.9 million users.

Domestic Bliss: A Good Deed Gone Viral in the U.S.

In a country with some degree of stability and few restrictions placed on our use of Facebook, the types of stories that resonate with users and gain traction in the U.S. can vary, from police protests and PETA-inspired, animal-welfare victories to memorials and tributes inspired by the loss of loved ones, such as the death of actor Robin Williams, among others.

One of my very favorite examples of Facebook's power to inspire on a collective level caught my attention one Saturday morning at my favorite café. I was enjoying my usual latte and scrolling through Facebook when a post from Ellen DeGeneres caught my attention. It contained a video of twenty-one-year-old Sarah Hoidahl's appearance on Ellen's TV show. Sarah was waiting tables at a restaurant when she overheard two members of the National Guard talking about their economic struggles during the U.S. government shutdown and decided to help them out. A single mother living at her mother's home, Sarah was also struggling to make ends meet; however, she felt moved by the soldiers and decided to pay for their lunch. She wrote them a note, which Ellen DeGeneres read to the studio audience:

> *Thanks to the government shutdown, the people like you that protect this country are not getting paid. However, I still am. Lunch is on me! Thank you for serving ladies! Have a great day!*
>
> *—Sarah*

The servicewomen posted Sarah's note on Facebook, which ended up going viral and catching Ellen's attention. At first, Ellen jokingly decided to pay this woman back the $27.75 of the total bill, but then later gave Sarah a 50-inch Insignia television (so that Sarah could watch Ellen's show), followed by a check for $10,000.00 to award Sarah for her act of kindness. The audience went wild with applause.

After watching this video I thought about all the nice people who perform random acts of kindness and, on that day, I felt particularly inspired to commit a random act of kindness myself. Perhaps I too could pay for someone's lunch or offer someone a free trip on the Chicago "L" train. The video had triggered a positive feeling of hope for humanity and led me to want to do my part, as small as it may be, to make the world a better place.

Later that day I found my opportunity when a woman in front of me at the deli forgot her wallet—*Lunch is on me*, I said.

'KNOWING IS HALF THE BATTLE'

Countless positive examples of Facebook's influence exist on a global and domestic level, but these effects can be difficult to focus on when we're suffering through the same medium (Facebook) on an individual level. We've seen firsthand through the case histories presented in this book that people are suffering the effects of Facebook, some people more predisposed to extreme behaviors than others. And because research is still underway, we can't know Facebook's true effect on our emotions, relationships and lives; however, we do know some things, and educating ourselves puts us in the best position to self-regulate our behavior and regain balance when we need it.

'INVASION OF THE BODY SNATCHERS'

With the increase of time spent online and the advent of social media our very brain chemistry is changing in ways we haven't even begun to understand. We are just now beginning to see daily articles on Facebook and research related to the affect social media and time spent online has on our physiology. They may be scarce, but the findings are surfacing. Facebook and other online activities have a significant impact on our bodies. We've mentioned in the course of this book how Facebook can have an affect on the production of our hormones. This is significant because hormones govern and regulate our bodily functions. For example, when we talk about ourselves, which we are doing more than ever through Facebook, Oxytocin is released. Oxytocin is also released when we get "likes" on our posts. And these are just a couple of the effects we know of without fully understanding how these and other hormonal changes impact our lives.

Physiologically we can change for countless reasons, but it's important not to underestimate the influence technology has on our physiology and psychology, as it's another way to maintain control over our emotions, lives and relationships. Staring at your computer screen for an extended period can alter your body's circadian rhythms and throw off your sleep cycle. Now it's being suggested that technology is shortening our attention spans and altering our neural pathways. More than ever before, people are suffering eyesight problems, headaches, fatigue, back and neck pain, all of which can be associated with online activities and significantly impact our lives.

Take a moment and think of how social media and technology have affected you over time. What have you noticed in your physical body? Do you find it difficult to concentrate on what the person in front of you is saying?

Privacy: *Mi Casa Es Tu Casa*

Another influence we should be mindful of is the corporate influence. Facebook is a company designed to make its shareholders and employees money. What started out as a way for Ivy League university students to stay connected quickly morphed into a billion dollar company whose main source of revenue is advertising and also requires you to forfeit some degree of privacy to participate. A political science professor I once met suggested that we can't even conceptualize what a life outside of advertising's influence is; it has a stronghold over our lives from infancy. Through Facebook's early design as "Facemash," the founders may have had a much simpler and benign vision for the social-media network, but things have changed since then, and not necessarily in users' favor. In forfeiting some privacy to participate, our emotions, relationships and lives have become subject to experiment.

A controversial research study published by the Proceedings of the National Academy of Sciences revealed that Facebook for one week in January 2012 worked with Cornell University and the University of California-San Francisco (UCSF) to conduct a social experiment on nearly 700,000 users to test their emotional reactions to pieces of content. The users weren't notified of their participation and unknowingly helped the researchers learn that people who read fewer positive words were prone to share more negative posts, while the reverse occurred when Facebook users were exposed to fewer negative sentiments. Although this information-gathering practice wasn't illegal due to Facebook users signing away many privacy rights when they agreed to participate on the social platform, this sneaky move on Facebook's part angered many of its users and makes the point that our emotions, relationships and lives are being unduly influenced by Facebook.

Take a moment and think about this: Can you say with any degree of certainty that you didn't lose a friendship, fight with a spouse, or experience some other life difficulty as a result of Facebook experimenting with your emotions?

WELL, AM I? AM I MY PROFILE PIC?

Of the many ways in which Facebook use affects our emotions, relationships, and lives, the most problematic to me is its effect on our self-expression and, thus, our self-esteem. From early on in our development we form our self-concept by projecting our reality outside of ourselves. We learn who we are and what we believe in through our interactions with others. We often compare our understanding of the world against others' understanding of the world. This is why, for most of us, our teenage years were so painful, because we were establishing our independence and self-identity and yet looking to our peers for validation. When we seek validation through Facebook, the effect on our psychology is similar to how it was when we sought validation while developing our self-concept.

Self-identity is not a static concept. Many of us struggle throughout our lives, making adjustments in our self-concept or identity, whether we are still in a struggle to establish our independence and sense of self, independent of our parents, or realizing through life experience that we've changed as a person. Facebook and social media are a new and lasting landscape in which Digital Natives are developing their sense of identity and earlier generations such as Gen X and Baby Boomers are having their identities mirrored back to them. The damage to self-identity lies in self-editing, seeking validation and reinventing ourselves as someone other than who we represent in the physical world, which can create anxiety from holding two conflicting ways we perceive our world.

Self-expression is an important part of happiness, and we now have an opportunity to express ourselves like never before. Self-expression can bring about a lot of fulfillment, and help bolster self-identity. Through Facebook and other social media, however, we have opened ourselves up to a larger community and, thus, critics. Facebook's features have made it easy for us to get pulled into a cycle of self-editing and validation seeking, which is undermining our self-identity and creating problems of self-worth. At an extreme level, many of us are defining our self-worth by the feedback we receive on the things we share, including heavily edited photos of ourselves. When we engage our self-editor we open ourselves up to self-doubt and lose pure expression. We can easily lose sight of who we are out of fear of how we appear to others. When our self-editor enters, and we edit ourselves to reflect who we think we should be, we can end up vastly misrepresented and lose sight of our true persona, which can lead to low self-esteem, depression and an overreliance on other people's opinions.

Take a moment to think about this question: Are you your profile pic? You are the only person who can answer this question.

Where Do We Go From Here?

The Internet and social media have made our lives easier and more enjoyable in many ways, but they were never meant to replace life. In order to find balance we need to reprioritize how we're spending our time. It's become the norm for people to look down with their gaze fixed on their smartphones. I challenge you to count the number of times a day that you use, or even just touch, your smartphone. If you're like me, you've learned to rely on it far more than you realize. And as we've seen through this book, our over focus on social media and technology is changing us.

Regardless of where we land on the scale of technology, social media or Facebook overuse, many of us can identify with the need to recalibrate once in a while, to pull back from our online time or virtual lives. We may recognize this need because we aren't sleeping well, we feel we've become addicted and are putting our relationships or job at risk. Or maybe we're turning into someone we don't recognize. Taking a step back, and being creative or establishing discipline in how we approach social media and technology, can bring natural balance back to our lives. Doing something as simple as sitting in silence can restore body processes we rely on to remain in good health.

The list on the next page gives obvious examples to help balance us in our use of social media and technology, but there are also other creative approaches. For example, these days, when we see something beautiful, hear something funny, or have an inspiring thought, our first impulse is to "share" the experience on Facebook. But what if you just took a few extra moments to appreciate that thought or city skyline or beautiful dish of food before automatically recording it?

If you compliment a friend on Facebook, make it a point to compliment someone in real life—that same day. If something inspires you online, don't just click the "like" or "share" buttons; remember that you have the power to act. Positive interactions on Facebook don't have to remain there. We gain so much information and awareness from social media, why not apply it to everyday life? If you come across something disturbing, do something to improve or change the situation.

10 Ways to Unhook from Social Media

1. Log out of Facebook when you're done posting.

2. Turn off any push notifications on your laptop or phone.

3. Put your computer to sleep and place it with your phone in another room before bed.

4. Give the person in front of you your undivided attention and demand the same.

5. Take a bath. But leave your hardware behind!

6. Limit yourself to checking Facebook three times a day for half an hour.

7. Take a weekly trip to a place with no cellphone reception.

8. Have a basket in the dining room or kitchen for phones during mealtime.

9. Don't use any form of technology after 9 p.m. at night.

10. Make it a point to spend an equal amount of time maintaining your offline friendships.

We are much more satisfied in life, and less inclined to need constant stimulation, when we take the time to appreciate what's happening around us. Be more mindful of your surroundings through your senses—sight, hearing, touch, smell, and taste. Let your eyes take in a beautiful painting. Take the time to enjoy the banter of friends over a nice dinner. Relish that piece of European chocolate, appreciate a fine wine, notice the scents and sounds of a crowded marketplace, embrace the warmth of your lover's arms a little longer.

Maintaining balance in our lives is key to happiness, and pulling back from technology is one way to help restore balance.

There Is No Hook

Facebook is a significant part of our lives, and as the No. 1 leading social-media website, it will likely be a long-term player in our lives and the lives of our children and families. While it's true that some users are predisposed to narcissistic and addictive tendencies, or can be categorized as one of the "manipulators" that we've studied in the book, for the majority of us, *there is no hook*. That's right. *There is no hook*. No one is forcing us to engage with one another through brief digital exchanges. We *do* have a choice. But part of making wise choices lies in understanding what we're up against.

Even if we've educated ourselves and understand the big picture implications of how Facebook and other social media affect us (by reading this book), we may still be at a disadvantage. In other words, half the battle may not be won in the knowing. It may be impossible to truly know and understand the effects of technology and corporate influence over our emotions, relationships, and lives. We may understand their effects in a general sense, but much about our lives has changed with the advent of technology and social media, and their influence over our lives is happening at a micro level.

The primary purpose of Facebook and other social media should be one of positive personal and social growth. Such growth is possible by first taking a moment to understand how we view others and ourselves. What we do and how we behave on Facebook expresses and reinforces this view, and if we approach each other with the understanding that we all have the same basic need to thrive—to get through our day by maximizing the things that make us happy and minimizing the things that don't—then Facebook becomes a powerful tool for personal and social growth. We have a personal responsibility to treat ourselves with dignity and respect, and a social responsibility to treat others in the same way. And when we behave responsibly, with purpose, and with others' and our own best interests in mind, we not only lift those around us but also ourselves.

With education and self-perception come the ability to self-regulate our actions, or to seek help if we are unable to moderate our behavior. We have the power within us to mindfully enjoy the features that new technology and social-media interactions afford us, while remaining genuinely true to ourselves and deeply connected to each other.

I am hopeful that any insight gained from this book will lead us to consciously choose to spend the majority of our time engaged and connected with one another, while assigning Facebook interactions a passive influence in our lives—second to the more meaningful experiences that can only be found offline.

appendix

Case Studies

I made a rule that everyone has to put their phones in a basket when we're at the dinner table. At first everyone stared at the basket while we ate. There's a sign of a problem. After a while, it got better. And now we don't even think about it.

Update Status

PRIVACY

Monique, 43
Charleston, South Carolina

I caught my kid posting about getting a blowjob on that thing. I almost killed him. I sat his ass down and told him that if he ever posts some crap like that again, he'll be sorry. And guess what? That stupid little girl liked that post! Can you believe it? No self-respect. Kids are crazy. Everything is out there. You're not supposed to share stuff like that. Teens don't know what they're doing on Facebook or the other sites they're on. They can get hurt or embarrassed or worse.

🖒 Like   💬 Comment   ➡ Share

Anabella, 24
Indianapolis, Indiana

The most upsetting thing I've seen on Facebook is when I checked my News Feed and saw a pic of a baby in a coffin. My friend seriously took a pic of her dead three-month-old daughter in a coffin and posted it—on Facebook. Who the hell does that? I was mortified. It's awful and wrong. I don't want to see that and I don't think she should have forced other people to see that either... Hell no, I didn't comment on the pic, but a lot of people did. I think that's just crazy.

🖒 Like   💬 Comment   ➡ Share

Betsy, 33
Chicago, Illinois

I got fired from my job because of Facebook. I received FMLA [Family Medical Leave Act] for depression and severe back pain, but I made the mistake of posting photos of myself sunbathing and partying in Cancun, Mexico. I forgot that I had friended my boss on Facebook. He posted a comment on my sunbathing pic: I'm so glad you're feeling better, Betsy. Let's have a little chat when you return home from 'the hospital.' When I got back to work, he fired me. I knew what I was doing I guess, but I really wanted people to see me sunbathing.

🖒 Like   💬 Comment   ➡ Share

Abdel, 35
Newark, New Jersey

I don't want anyone to know anything about me on Facebook. I told my wife not to say we are married on there. People are crazy on that thing, and the less I'm involved the better. My wife's on Facebook and that's fine, but I don't think that it's right for everyone to be in your business. People need better things to do with their life than to snoop around everyone else's business. None of it makes any sense to me.

    👍 Like   💬 Comment   �که Share

Miguel, 35
Santa Fe, New Mexico

My wife wanted to take selfies of us kissing at the altar during our wedding ceremony. I told her she was crazy. She didn't see anything wrong with it. Doesn't that seem crazy to you? I mean, we can't even enjoy our wedding ceremony without her posting a Facebook update. I wouldn't let her do it. She's still nagging me about not letting her take a wedding-kiss selfie.

    👍 Like   💬 Comment   ➤ Share

Carolina, 28
Albany, New York

I told one of my friends something embarrassing that happened to me in high school. It was pretty bad. She shared it on Facebook and wrote: *Guess what happened to Carolina?* That was supposed to be private. I don't know if I was more angry, hurt or confused. People joked about it and I tried to say that it didn't bother me, but it did. I think she shared it on her wall so that she could make fun of me. I haven't spoken to her since.

    👍 Like   💬 Comment   ➤ Share

Josephine, 37

Madison, Wisconsin

I refuse to post any pics of my children on Facebook. I'm very strict about that. I was a criminal history major and I know that there are a lot of sick people out there. I don't know how mothers are okay with sharing photos of their children like that. When my kids are older they can share what they want but as far as I'm concerned, I think it's wrong and it's dangerous.

◌ Like   💬 Comment   ➡ Share

Chris, 23

Washington D.C.

The craziest thing of all time is when you see someone going through a break-up on Facebook. You get a play-by-play. I've texted my friends, *You've got to check this out!* People insult each other, blast each other, post things to make their ex jealous, then regret it and post love songs or quotes to try to get them back. Everyone gets to see just how dysfunctional relationships really are.

◌ Like   💬 Comment   ➡ Share

Naima, 28

New York City, New York

I was trying on my wedding dress and my best friend was taking photos for me. After we left I found out that she posted them to Facebook. I was so mad at her. My fiancé saw them and liked a few of them. I didn't want him to see my wedding dress. She should've known that. She said that she just posted them without thinking about it. I didn't talk to her for a week I was so angry. She's not a dumb person. She knew better. I'll never figure out why she did that. It's going to take me a while to forgive her.

◌ Like   💬 Comment   ➡ Share

Update Status

ADDICTION

Antoinette, 32
Crystal Lake, Illinois

I love Facebook. I have a serious Facebook problem, but at least I can admit it. Ha! I'm on it all the time. I'll be on it at 3 a.m. when I can't sleep. Sometimes when I'm checking the News Feed I won't fall asleep at all. My friends would probably tell you that I'm on it all day, and it's true. If it didn't exist, I'd get so much work done. I don't know why I need to check it—I just do. I just know that there's no way I could go a whole day without Facebook. I think I'd die.

Like   Comment   Share

Stephen, 41
Charleston, West Virginia

I'm really into this girl right now. We recently met again after not seeing each other for a long time. It's because of her that I got addicted to Facebook. Sure, blame it on Facebook. I'm kidding, but I was checking that thing like twice an hour to see if she posted anything. I got worried that she could see that I was checking out her wall too much. Can people see that? She started posting pics with her and this dude. Why is she doing that? Do you think that she's dating him or do you think she's trying to make me jealous? If I like that pic, I think she'll see that I'm confident and I don't care. What's better, liking these pics or ignoring them?

Like   Comment   Share

Angie, 27
Detroit, Michigan

I'd rather have Facebook than sex. I can imagine a week without sex, but I can't imagine a week without Facebook. I need it. Without it I just can't imagine what my life would be like.

Like   Comment   Share

Brian, 19

Chicago, Illinois

I developed tendonitis in my thumbs because I text and check Facebook on my phone so much. Sad right? Now I can't text without them hurting. I'm getting a brace soon, so I'll have to use the voice command maybe? I know what you're going to tell me, how about I take a break? Maybe, we'll see. Or, maybe I can get someone to scroll through Facebook for me, lol.

 Like   Comment   → Share

Dee, 35

Salt Lake City, Utah

I had to take a break from Facebook. I just needed to turn if off for a while. It's too much sometimes. There were times when I wasn't even looking for anything or at anything. I was just scrolling like a robot. You get programmed to check it all day. I didn't even know why I was on there anymore. It didn't make me happy. It's the same stories over and over. It's not doing anything for me anymore.

Like   Comment   → Share

Barb, 24

Augusta, Maine

I broke up with my boyfriend because he would never put his damn smartphone down. We'd be on a date or in his car, and he'd drive and text or Facebook. It drove me crazy. I told him to stop, his mother told him to stop—a lot of people told him to stop. He was looking at his phone all the time. He never looked up at me. We'd be at dinner, and he'd keep checking his phone. He'd try sometimes, but even when we were talking, I could see he was nervous. He'd stare at his phone. I can't date that. No one can date don't care. What's better, liking these pics or ignoring them?

Like   Comment   → Share

Bruce, 44

Saint Paul, Minnesota

I'm going to tell you something. People are crazy on Facebook. And I mean crazy. They need to tell you every damn thing that's going through their head. I went to the grocery story. Really? We needed to know that? Then they fight and attack each other on there. Then they break-up because someone got caught cheating with this chick or someone got jealous over nothing. It's bullshit. You know what? People are crazy—that's a given. But Facebook makes people crazy AND stupid.

Like   Comment   Share

Update Status

V<small>ALIDATION</small>

Nolan, 29
Atlanta, Georgia

I like reading stories on Facebook about stupid criminals that get caught because they posted their crime on Facebook. It's too funny. I mean, they're so excited to take a selfie in front of the money they just stole and then post it on Facebook, that they don't even think that someone may report it to the police. Morons.

        Like   Comment   Share

Margaret, 40
Malibu, California

My husband got so angry with me because he said he had tons of friends like his posts, and I was the only one who didn't like them. He said that of all people, I should like his posts. I told him that no one really cares about his posts, they're liking them out of friendship obligation, but no one really cares. Everyone can see that he likes to try to impress people. I don't like them because they seem so sad to me. We fought over Facebook every night. I left him three months ago. I just couldn't take it anymore.

        Like   Comment   Share

Andre, 26
Monterey, California

My friends tease me about my gym posts. Every time I'm there I post it on my wall. I do this 'cause I want people to know that I'm there and that I'm committed. It makes me feel good knowing that they know that I'm an athlete. A couple of times I went to the gym but forgot to check-in. I was so pissed. It doesn't really count unless my Facebook friends are watching. It's simple really. If people like what I say or do then it makes me feel good. If they don't like my stuff I get nervous—like uneasy. I need them to like that I'm at the gym 'cause it's a part of me.

        Like   Comment   Share

Alex, 43
Grand Rapids, Michigan

I got to be an extra on a film, so I kept posting about it. I thought it was cool so I wanted everyone to know where I was. One of my friends calls me to tell me that I look needy, like I'm trying to get attention because I keep posting about it for three days straight. But I don't think I was seeking attention. I was just sharing a cool experience. No one else said I was seeking attention, so he's wrong.

    👍 Like   💬 Comment   ➡ Share

Anna, 30
Toledo, Ohio

I gained weight after my pregnancy so I stopped posting on Facebook. Yes, I know, I'm supposed to gain weight, and everyone gains weight, but I don't want to post photos of me like this for everyone. If my friends visit me, that's okay. I don't think about it. But on Facebook, I get so depressed even trying to choose a pic where I look decent or okay. I compare these to the way I used to look—flat stomach, great abs—and I feel sad. I don't like the way that I look and I don't want people to see me like this. I'll drop some weight, then I'll post photos again. Until then, I'll just talk about the baby. This takes the attention away from me.

    👍 Like   💬 Comment   ➡ Share

Suma, 44
Tempe, Arizona

I used to be a triathlete before I got injured. I loved posting pics of me running or biking. Now I don't like Facebook, because I feel like I'm not a part of it, you know? Swimming, biking, and running were a big part of my life and all my friends saw this on Facebook. Now all they do is ask me how I'm doing and I hate that. I want them to see me on there as the bad-ass that I used to be, not someone that needs them to check-in on me. I was going to post one of my old pics, but then I thought, that's just sad. I don't even bother with Facebook anymore.

    👍 Like   💬 Comment   ➡ Share

Update Status

R<small>ELATIONSHIPS</small>

Morgan, 41
Washington, Illinois

I check my ex's Facebook religiously. She had 376 friends yesterday but now she has 378. I'm trying not to drive myself crazy wondering which guy she's talking to next but I can't help it. Who is she talking to now? It's right in my face—right there. There's no way you're not going to look. I try not to obsess like this. I know it's stupid but it's hard. It's confusing. It messes with you. What should I worry about or forget? What's real and what's not?

 Like   Comment   → Share

Ted, 35
Bridgeport, Connecticut

She shares every damn thing on Facebook. If we get into it, all of our friends know about it. Then they all comment shit like 'You're too good for him anyway.' Facebook is the place where all these girls pump up each other's egos—'You look so hot,' and slam men. When we make up, they have the nerve to be nice to me to my face, after all the crap they shared online. I told my girlfriend to quit posting about our fights, but she said Facebook is like therapy. If she keeps this up, it's over.

↻ Like   Comment   → Share

Emily, 27
Austin, Texas

I thought I really liked him, but then I started noticing how he was treating other people on Facebook. He slammed his friends, made nasty comments on my friends' walls. When I confronted him he'd say he was just kidding, but it's obvious that he doesn't respect women. Plus, he'd post updates showing the charity work he did and how he's supposedly a humble person. But posts like that reveal your true colors. It means you just want likes and praise for displaying how humble you are. Facebook: it's the ultimate way to display your hypocrisy.

↻ Like   Comment   → Share

Leslie, 40
Denver, Colorado

I found out that my husband was cheating on me on Facebook. He changed his Relationship Status from Married to Single and then started posting pictures of himself with someone else. I found out about this through my friends who told me to check my Facebook wall. That was the cruelest, most horrible thing anyone has ever done to me. When I finally got ahold of him, he told me that he was sorry, and that he did let me know—on Facebook.

    👍 Like   💬 Comment   ➦ Share

George, 28
Orlando, Florida

You can't break up with someone and still be their Facebook friend. You just can't. It's too easy to stalk them, because you miss them and you still want them to be a part of your life—even if it's just to see what they're up to. Then you think before everything you post on your own wall, 'Will she see this? Will she miss me if I post this?' You become a slave to what they might think. It's too much anxiety for one person to handle. When you break up with someone, just break up—in every way.

    👍 Like   💬 Comment   ➦ Share

Janie, 34
London, England

I had to get a restraining order on my ex-boyfriend because he got violent. Every time I liked a pic from one of my friends, we had an argument. He'd ask me about every single guy I friended. He demanded that I unfriend my good guy friends. He wanted my Facebook password. I said no. Then he said that there should be nothing between us—so I should give up my password. I wouldn't, and he smashed my car windows. I left him three weeks ago. People tell me that he keeps posting about how much he loves me on his own wall. I'm going to have to get another restraining order to get him to stop that.

    👍 Like   💬 Comment   ➦ Share

Linda, 46

Boise, Idaho

I got revenge on my husband for cheating on me. He took pictures with this girl at his high school reunion. I saw this on Facebook. They were just too close, you know? Then I found texts between them on his cellphone. We shared a Facebook account at that time so I got on there and said that this girl's a whore and that I'm cheating on my wife of 12 years with a whore. I tagged her in each comment so her family could see; so her children could see the kind of woman she was. I'd do it again if he cheated with her again. Yes, I needed to post this on Facebook. I wanted the world to see what they were doing. My husband got so mad at me and said that he wasn't cheating, that they're just friends and I embarrassed both of them over nothing. Oh really? Then why did you take a pic with her and post it on your wall?

Like   Comment   Share

Robin, 32

Cork, Ireland

Staying on Facebook made me crazy. I became absorbed in it and couldn't stop. I became obsessed with my ex and with what he was doing. He posted things to try to make me jealous or to get a rise out of me. People kept telling me to ignore it or to block him, because I'm just giving him what he wants—attention. But there's something about Facebook that keeps you coming back. It's like you lose who you are and take on everything that this person is posting. I couldn't stop no matter how desperately I tried to. His posts turned me into an obsessive crazy woman. And before Facebook this was never me. I never chased or followed anyone. When it was over, it was over, but Facebook keeps you attached. It's like cocaine. Your desperation for a normal life doesn't hold water to Facebook's hold on you. It keeps you connected to someone even if they're the worst possible person to have in your life.

Like   Comment   Share

Update Status

IDENTITY

Andy, 47
Louisville, Kentucky

I hate it when people are fake on Facebook. They actually think they're fooling people. Everyone is showing only the good things. Why don't people post the normal things? That's what I do. I don't try to impress anyone because I'm me. I don't try to be anyone else. A lot of people on Facebook, all they do is post their accomplishments, their gyms, and them being great moms or dads. No one is okay with being themselves anymore. I don't know who people are anymore.

☺ Like   💬 Comment   ➙ Share

Nancy, 38
Cincinnati, Ohio

Before Facebook, I never cared so much about how I looked. I mean, I did a little, just like other people. But I hate being tagged on a bad pic. It's like a mask that you can't take off and your bad pic is forced up there on a wall, until you can call your friend and ask them to untag you. I know I'm not the only one who feels this way. Everyone wants to look good on Facebook—that's why we do selfies—so that we can control how we look.

☺ Like   💬 Comment   ➙ Share

Beverly, 19
Fort Worth, Texas

I do something different than most people. I post silly stuff and make fun out of myself all the time. I figure why not? Life isn't perfect, and I'm not either. I'm hoping other people will do the same. My mom says that if you can't laugh at yourself, you're going to suffer a lot in life. I do the dumbest things, and I'm a total goofball all the time. That's how my friends know me, and that's how I express myself.

☺ Like   💬 Comment   ➙ Share

Cassandra, 36
Bridgeport, Connecticut

I didn't want to friend some of the people I went to college with. I got really fat and they still look normal. If I friend them, then they're going to see how fat I got. I see their profiles. They have perfect lives, with perfect marriages, and perfect kids, and I'm just me. They look so happy all the time. They keep asking me how I am, but I know what they're really thinking. They're looking at my photos and wondering how I got so fat. I hate Facebook.

    Like   Comment   Share

Ruben, 20
San Jose, Costa Rica

My friends joke with me about my thesis-level posts on Facebook. I enjoy sharing my reviews of books and foreign movies, but what I really like posting are my existential ideas about life. It's true that I post too many times each day. I'm a philosophy major. I'm pretty good at sharing the deeper stuff. Very few people understand how I think when I express myself face-to-face, so I turn to Facebook for understanding. I have over 1,200 friends. A lot of people appreciate my thoughts. Most people like to read about the meaning of life and I provide that for them—some insight into what we're all doing here and the point of it all. It's who I am When I see how many likes I get for my posts, I feel good about myself and know that I picked the right field of study.

    Like   Comment   Share

Update Status

F RIENDSHIP

Emily, 27
Austin, Texas

One of our friends, Jessica, likes to poke fun at me on Facebook.
We all notice it. Whenever I post something inspirational, she makes
a joke about it. When I post something funny, she comments that
what I wrote is lame. It got really annoying. I told her to quit it, but
she told people that I can't take a joke. Since when are insults and
jabs considered funny? I got sick of it, so I unfriended her, and then
she told our friends that I'm immature. Um, no sweetheart, setting
boundaries with mean friends doesn't make you immature. Acting like
a brat when you don't get your way is what makes you immature.

 Like   Comment   ↱ Share

Michael, 32
Anchorage, Alaska

I'm Italian, which means that I have a hairy back. So I use this product
that you can just wash off that removes your hair. I used a towel
in the bathroom and I accidentally left the towel that I used on the
bathroom floor. My roommate decided to take a picture of the towel
and posted it on Facebook with a comment, 'Just when you think you
can't get more grossed out.' Granted, I left the towel there by accident.
I shouldn't have, but she didn't get why I'd be upset that she posted
that picture. She said it was no big deal because she didn't tag me on
that pic, but we have mutual friends who know that I'm her roommate.
I told her I was angry about what she did, but she still doesn't get it.

△ Like   🗨 Comment   ↱ Share

Juliet, 45

Madison, Wisconsin

My sister doesn't speak to me anymore because I didn't like enough pictures of my niece on her Facebook wall. Forget that I was always there for her, that I supported her, defended her, protected her, and was there for her in her times of need. Forget all that, right? The only thing that seems to matter anymore is that I posted something on Facebook that didn't sit well with her, or that I didn't like enough pics of my niece. This is ridiculous. I spend one week taking a Facebook break and all of a sudden I'm a horrible sister. I did nothing to deserve the way she's treating me. After the way she cut me off like that, our relationship will never be the same again.

       👍 Like   💬 Comment   ➡ Share

Max, 40

Chicago, Illinois

A buddy of mine started dating this model, and he thinks she's beautiful, but she's not nearly as hot as the model that I'm dating. So he started posting pics all over his wall of her. It was like way too much. She's honestly not as hot as he thinks she is. So I posted a joke on his wall, 'Just share another pic of your average girlfriend please!' I was just messing with him because he shows off his girlfriend in everyone's face. He got pissed. I guess that's insecurity for you.

       👍 Like   💬 Comment   ➡ Share

Teens

Laurie, 16

Montgomery, Alabama

This boy in my class started flirting with me on Facebook and Twitter. I didn't think he liked me because he hangs with a different crowd, and I didn't think I was his type. But he kept posting poems and stuff like that all over my wall so we started texting. We met up a couple of times. I really liked him. We ended up hooking up and everyone found out about it. Turns out he was posting that stuff so that everyone could laugh at me—like a show. Afterwards, people kept posting on my wall stuff like, 'He doesn't like having sex with fat chicks you dumbass,' and 'you're such as stupid bitch for thinking he would seriously be into you.' I cried for weeks and didn't want to go to school. My parents sent me to a therapist 'cause I wouldn't eat. I know I'm fat. Everyone knows I'm fat.

⟳ Like   💬 Comment   ↪ Share

Aleshia, 16

Boston, Massachusetts

My mother tried to pay me $500.00 to leave Facebook for six months until my grades improved. For $500.00, of course I deactivated my account! Who wouldn't do that? I stayed off of it for a week, but I couldn't stay off of it for long. I reactivated my account and paid back what I didn't already spend on clothes, and had to pay the rest back slowly from my paycheck. My mom thought this was all, like, so hilarious, and she posted what happened on her Facebook wall. That wasn't fair, right? So now I use Twitter.

⟳ Like   💬 Comment   ↪ Share

School Principal
Chicago, Illinois

These students don't understand that when you post something on Twitter, Facebook, or Snapchat it's on there forever. We hear about this stuff all the time. Parents come in complaining because their son or daughter witnessed something traumatic on Facebook and they want administration to punish the other student. It's not always that easy. There's a fine line between how much school officials can intervene. Sometimes we encourage parents to call the police. We cannot always make students stop bullying each other online. There's no way to control it. We do what we can, but parents need to understand that it is also their obligation to monitor their child's online behavior. They cannot keep pointing the finger at us.

ᗜ Like   💬 Comment   ➡ Share

Brenda, 15
Releigh, North Carolina

This girl cut on herself and posted it to her Facebook wall. At first I thought it was a joke, but then people started commenting on it. It was real. My friend called her mom to tell her about what was happening. They don't talk anymore because she got sent to a hospital for a month. Now everyone looks at her weird in the hallway.

ᗜ Like   💬 Comment   ➡ Share

Hazel, 40
Bend, Oregon

My son's grades are suffering because of Facebook. We've tried everything. We took the computer out of his room and put it downstairs, we try to engage him but he won't talk to us. We try to spend more time as a family. I've hired a tutor and it helps, but whenever he's not busy, he's on Facebook all day. That's where all his friends are now. They don't seem to even go out as much as they used to. They text and stare at a computer. I've spoken about this with other parents, at school meetings and it's a serious problem.

ᗜ Like   💬 Comment   ➡ Share

Ben, 16

Corpus Christi, Texas

There's this girl at school who started posting on Facebook about how she's a lesbian. She keeps posting this stuff and adding the rainbow pics. You know which ones, right? So now she and this other girl are dating, and they hold hands in the hallway. On Facebook there are a ton of pics of them making out. It got me all freaked out about it. Other people started sharing the pics on their walls, too. I don't mind that she's gay. My mom has friends that are gay. It's cool, but keep that shit to yourself, you know?

Like — Comment — Share

Celia, 15

Chicago, Illinois

This girl at school has bulimia. She posts about it on Facebook. Her mom's on there so she knows too. She talks about what the doctor says, what the hospital says, what her therapist says, you know? She takes like these pics of herself showing off how skinny she is. It's so ugly. People post how they care about her and how they're praying for her. My dad said to be friendly but to not get involved, 'cause a lot of people are already involved.

Like — Comment — Share

Update Status

EMOTIONAL MANIPULATORS

Private Investigator
Chicago, Illinois

I'm a private investigator. I can't tell you how many times we get involved in cases with Facebook cheating or stalking. Everyone's an investigator now. People tell me that they've spent days or weeks on Facebook trying to catch their wife or husband cheating. Most of the time if someone is cheating, they themselves actually make our work easier. All we have to do is check their Facebook—people usually end up giving themselves away. People think these things are private, but they're not. If you want to find something about someone you will, because everyone wants to share their story.

ﭼ Like   🗨 Comment   ➙ Share

Sandy, 34
Biloxi, Mississipi

Martyr: One of my friends constantly talks about how depressed she is. Why do people do that? I keep telling her, 'Go see a therapist. Go see a psychic. Go see a shaman. Go take a Percocet. Just go get some help already.' But no—she wants to post about it on Facebook, so that everyone can show her support and ask her over and over what's wrong. I'll tell you what's wrong—Sister Friend needs herself some Prozac. That's what's wrong.

ﭼ Like   🗨 Comment   ➙ Share

Martin, 25
São Paulo, Brazil

Seducer: Some of these girls are so stupid—they fall for anything. I like messing with them. I flirt with them online and they tell you everything. Sometimes I like to see how far I can go, how badly they'll let me treat them. They get pissed when you don't return a text, but don't like one of their pics, or ignore them on Facebook? They all go crazy. Then pretty soon, you'll get 'Why don't you love me?' Ha!

ﭼ Like   🗨 Comment   ➙ Share

Randy, 44
Chicago, Illinois

Seducer: The checking of her wall got so bad that I couldn't take it. One day I was on the 'L' platform, and I thought about jumping in front of the train. That's how bad it got, but something stopped me. That's when I knew that I'd enough. I needed to get off Facebook—for good this time. I had tried several times, but couldn't. Why? I don't know why. It's like a part of life, but not the kind of life that I wanted. The pain that I felt when I'd see her comments was horrible. Finally I decided that I wanted my life back, so I went through detox to get off of it. Once I left, and hung out with my buddies more, I came back to the old Randy. I slowly became happy again.

    ♡ Like   💬 Comment   ➜ Share

Leanne, 29
Orlando, Florida

Stalker: I had to cut him out of my life because he kept posting inappropriate things. We have mutual friends and it was really embarrassing. On Facebook, he kept posting how he loves me and that he can't live without me. I kept telling him to stop, but he wouldn't listen. I had to unfriend him, but this asshole started posting things for me to see on other people's walls! Then he started following me on Twitter, Pinterest, Instagram—all of them. Then the texting harassment started. I had to block him from my phone, my work phone, my email, and work email. He was acting all crazy. It got scary. Now people are telling me that he still talks about me on Facebook. Believe it or not, in person, I thought he was actually kind of normal, but on Facebook? Holy crap—he got crazy. Someone send this man to therapy.

    ♡ Like   💬 Comment   ➜ Share

Vincent, 39

Jackson, Mississippi

Stalker: After we broke up, I followed her every move. I knew she was telling me things—without really telling me things. Like when she checked-in to this bar—that was our place. She was trying to make me jealous. She was there with someone else, but she was really expressing how much she wanted to be with me. I spent hours scanning her wall, looking at the comments to her friends' walls. I needed to know what she was thinking and feeling at every moment, and Facebook made that possible. When someone likes a video or pic—you know what they're thinking even when they're not around you. You still have access to them.

👍 Like   💬 Comment   ➡ Share

Kim, 46

Frankfort, Kentucky

Narcissist: A friend of mine, well she's no longer a friend, completely ignored me on Facebook. She said that I cursed too much, and she didn't want to like my posts because they showed up on her News Feed or some shit like that. She didn't want people to know that she liked my cursing, so I unfriended her and cut her out. How are you going to like other people's posts and not like mine? That's crazy.

👍 Like   💬 Comment   ➡ Share

Cynthia, 31

Denver, Colorado

Narcissist: I post about going to the gym, about a new great outfit I'm wearing, new shoes, the best restaurants, the best fashion shows, and the best men. I have like 2000 friends—more than that. And they like everything I post. I don't have time for boring people so I unfriend a lot of people. They're so annoying. No one cares about you, okay? No one cares about your ugly husband. You're boring. Stop posting about your stupid job or your stupid family or your stupid life. No one cares. Seriously.

👍 Like   💬 Comment   ➡ Share

Darlene, 53

Seattle, Washington

Saboteur: One of my coworkers asked me to friend her on Facebook—
huge mistake. First she started liking every single thing I said or
posted on my wall. It was like she was following me. Then she started
posting these crazy comments like, 'I don't think you should have done
that' or 'That's not what a Christian woman is supposed to do.' I mean,
what? In person she acts nice to your face but on Facebook, she was
judging everything that I did, so I just hid my posts from her. When
I got to work she confronted me about it, and I told her why I did it.
After some time, she seemed normal again, so one day I decided to
post a group work photo of us and tagged her on it. She got angry
that I tagged her without her permission; she made a big deal and told
everyone about it at work. I had enough and just blocked her. Holy
shit—then she really went nuts. She gossiped about me, told my boss
lies about me and some other stuff. Then she yelled at me from across
the room, 'I didn't know you were so immature Darlene! You actually
unfriended me on Facebook!' She said this in front of everyone. You
would have thought I burned her house down or something.

ò Like   Comment   → Share

Sam, 44

Lansing, Michigan

Saboteur: I've been dealing with pain for a while. Mostly migraine
headaches and neck pain. My wife posted something about it on
her Facebook. My sister-in-law wrote on there, 'I wish I had time to
get sick, but I have too many obligations.' The more I sat with it the
more pissed I got. Who are you to tell me that I'm not in pain? You
don't know anything about how I feel. The next night our families got
together for a cookout, and I talked to her and told her that what she
said made me upset. She apologized; she said that she didn't mean
it—fine. It was done. When she got home, she posted, 'It's difficult
to cope with people who are so sensitive and emotionally weak.' I
knew that was about me. If she wants to act like a bitch, then go right
ahead, but she should at least have the guts to say it to my face.

ò Like   Comment   → Share

Acknowledgements

Much appreciation and gratitude are owed to those who helped me in the creation of this book. First and foremost, I would like to thank all those individuals who shared their Facebook experiences with me, and allowed me to quote them, with the intention of helping others who may be going through similar experiences.

It takes an astounding number of talented and gifted people to produce a book. On the top of that list is my amazing agent, Elizabeth K. Kracht, who has been on my writing journey from the beginning. Your guidance is invaluable, and your friendship is irreplaceable. I would like to thank Reputation Books and Mary C. Moore for picking my book out of Liz's slush pile and also guiding me through the fine-tuning process. I thank my copyeditor, Michelle Richter, for bringing out the best in my words,

I am indebted to Leslie V. Waite for assisting me in formulating the foundation of this book, for contributing her insight, and for helping me to find my voice. I would like to thank Dr. Stanley Tam for his countless hours of proofreading, suggestions, and not-so-gentle critiques that forced me to up my game. A special thank you to Annemarie Tracey for her assistance with networking research and to Jerry Cleaver for his excellent advice. I am very grateful to each of you for your contribution, support and friendship.

I thank my parents, Jose Mario and Rosa, my aunt Elena, my brother Mario, my sister-in-law Sarah, and my in-laws Betty, Michael, Christopher and Meredith, who supported and encouraged me in spite of all the time it took me away from them.

To my dear friends Silvia, Derrick, Niranjani, Matt, Michelle, Bob, Lidia, and Jill for reading sections of the manuscript and offering advice, for always having my back and for cheering me on when I needed it most. Thanks to all my friends who offered support and feedback along the way. I beg forgiveness of all those who have been with me over the course of this journey and whose names I have failed to mention.

Above all I want to thank Marc. These past few years have been tough on you, too, and I have much to thank you for. Thank you for always standing by me no matter what. Thank you for believing that the book is good, and that I am good, and that I was the person to write it. You are my inspiration and my love. Let's renew our vows in New Orleans okay? I'll wear one of my crazy hats.

About the Author

With insights drawn from over ten years of clinical experience, Dr. Suzana Flores shares her expertise in anxiety-related symptoms and how they manifest and present themselves through social media. Over the past three years, Dr. Flores has studied the Facebook phenomenon extensively, interviewing people from across the globe on their experiences with social-media interaction and how it has affected them. As social-media expert and commentator, she is sought after as a speaker and author on national and international newscasts, podcasts, radio and talk shows. She is a regular blog contributor and is regularly featured on *The Ron Kelly Show*. Dr. Flores holds a Doctorate in Clinical Psychology from Argosy University and a Master's in Counseling Psychology from Loyola University Chicago. She has served as Lead Clinician at Carolinas Healthcare System, Director of Counseling at the Illinois Institute of Art and Adjunct Professor at The Chicago School of Professional Psychology. She is a licensed clinical psychologist working in private practice. She lives in Chicago, Illinois.

Reputation Books